AF531679

ENCYCLOPAEDIA OF MOLECULAR BIOLOGY

ENCYCLOPAEDIA OF MOLECULAR BIOLOGY

(Vol. 1)

A. Tyagi

ANMOL PUBLICATIONS PVT. LTD.
NEW DELHI - 110 002 (INDIA)

ANMOL PUBLICATIONS PVT. LTD.
H.O.: 4374/4B, Ansari Road, Darya Ganj,
New Delhi-110 002 (India)
Ph.: 23261597, 23278000

B.O.: No. 1015, Ist Main Road, BSK IIIrd Stage
IIIrd Phase, IIIrd Block,
Bangalore - 560 085 (India)
Visit us at: www.anmolpublications.com

Encyclopaedia of Molecular Biology

© Reserved

First Published, 2006

ISBN 81-261-2677-9 (Set)

[Responsibility for the facts stated, opinions expressed, conclusions reached and plagiarism, if any, in these volumes is entirely that of the Author. The Publishers bear no responsibility for them, whatsoever.]

PRINTED IN INDIA

Published by J.L. Kumar for Anmol Publications Pvt. Ltd., New Delhi - 110 002 and Printed at Mehra Offset Press, Delhi.

Contents

Preface

Smallest fundamental unit of an element or a compound is called a molecule. The rapid progress and multi-faceted development of Science, in general; and of Biology, in particular, has ushered in an absolutely new era in the history of mankind, regarding the physio-chemical department of life.

There has been an urge to know more and more from the day one. It is due to this urge that man has been scaling ever new heights day-after-day, be it related to Science or Commerce or Politics and so on. Is this whole endeavour, which has proved to be a real boon for man, or in other words, the greatest gift, the twentieth century gave to humanity ? It really brought about a revolution in the whole human life. The emergence of Molecular Biology, as a branch of Science for study, could also be possible by virtue of this very present of the past.

The discipline is very vast and with the passage of time, big breakthroughs are in the offing. Feeling the pulse of the time, it deemed suitable, rather, necessary to bring out a comprehensive and compact encyclopaedia on this subject in order to cater to the maximum requirements of the lovers of Molecular Biology.

Comprising five volumes, this trivial attempt is expected to prove to be an asset for Science elite and students alike. Now, savour it molecule-by-molecule.

Editor

1

Gene Expression

When a 'virulent' bacteriophage (like T2, T4, SPO1, etc.) infects a host bacterium, it entirely depends on the host for its multiplication. It subverts the host's function and utilizes the host machinery for producing a large number of progeny phage particles. The bacterium cell undergoes lysis and dies to liberate a large number of these phage particles, which are each then ready to start another cycle by infecting new bacterial cells. This cycle is known as lytic cycle. However, there are other phages called 'temperate' phages which have a dual existence such that they may either perpetuate through a lytic cycle as above or may take the form of a 'prophage' by integration of their DNA with the DNA of bacterial cell. The latter condition is called lysogeny', which provides immunity against infection by further phage particles of the same type, so that there is only one copy of phage per bacterial cell. We know that this prophage may become free, turn virulent and then take the course of lytic cycle as in virulent phages.

When the phage DNA enters the bacterial cell leaving the protein shell outside, it directs the host cell for preferential replication of phage DNA. In doing so, the phage genes are expressed in a particular order. The lytic cycle is divided into (i) 'early infection' which covers the period from the time of infection to the start of replication of phage DNA and (ii) 'late infection' which includes the period from start of phage DNA replication to the final step of 'lysis'. During this life cycle of a phage (particularly λ phage), different genes of the phage genome are expressed in a regulated manner and form a cascade of gene expression, so that the product of one gene regulates the expression of the next gene. These aspects of the regulation of gene expression in phages will be discussed in this chapter.

Cascadal Regulation

In phages, the early, middle (or delayed early) and late genes are expressed in an order and a study of genetic map suggests that genes are arranged on the map in the order in which they are expressed. This, therefore, needs smallest number of switches and the regulation is exercised with a maximum economy. In the very early genes, the transcription apparatus of the host is utilised and the product is then used for initiating the transcription of the next class of genes (delayed early or middle genes). The transcription of these middle genes will be necessary for expression of late genes through a regulator, such that these successive controls (where expression of each set is necessary for the expression of the next set of genes) creates a cascade, in which a group of genes are turned on or off at a particular time.

Synchronous Sigma Elements

The regulation of gene expression through a cascade has been studied in phage SPO1 infecting Bacillus subtilis. It has been shown that for the expression of early genes of SPO1, bacterial RNA polymerase ($\alpha_2\beta\beta'\sigma^{55}$) is used, because the promoters of these early genes like the promoters of host genes can be recognized by this RNA polymerase holocnzyme. The consensus sequence in SPO1 at -35 and -10 base pairs resembles those of E. coli and B subtilis. For the expression of middle and late genes, however, expression of at least three regulator phage genes, namely genes 28, 33 and 34 is needed. The products of these genes are respectively called gp^{28}, gp^{33} and gp^{34}. The gene 28 is an early gene, whose expression is needed for the expression of middle genes. Similarly, genes 33 and 34 are middle genes whose expression is a prerequisite for the expression of late genes.

It has been shown that gp^{28} replaced σ^{55} in the holoenzyme $\alpha_2\beta\beta'\sigma^{55}$ and therefore gp^{28} is sometimes also referred to as o^{gp} This RNA polymerase with gp^{28} can then transcribe the middle genes. The next transition involves replacement of gp^{28} by gp^{33} and gp^{34}, so that RNA polymerase may now transcribe late genes (Fig.). Thus there are at least three sigma factors used in phages SPOl, namely σ^{55}, $\sigma^{gp^{28}}$, $\sigma^{gp^{33}\text{-}gp^{34}}$. We have seen the replacement of σ^{43} by σ^{37} and σ^{29} during bacterial sporulation, suggesting that selective

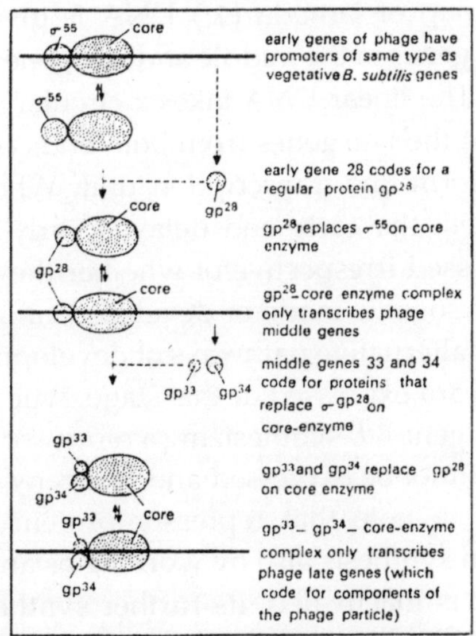

Involvement of several sigma factors, which change the specificity of host RNA polymerase for transcription of phage SPO1 genes.

transcription can be achieved by replacement of sigma factor in RNA polymerase holoenzyme. We do not clearly know about the factors which cause this replacement of one sigma factor by another, but there are different promoter sequences in different genes to be recognized by different sigma factors.

Effect of Lytic Cascade in Lambda (λ) Phage : Lambda (λ) bacteriophage exhibits one of the most interesting, but intricate cascade circuits for the two alternative pathways (lytic cycle or lysogeny), that it can follow after infection. The readers may be aware about the structure and the mode of infection of phage λ. The temperate phages like lambda can be classified as episomes, since these are dispensible and exist in free as well as integrated states. The phage can infect bacteria and multiply in the normal manner leading to lysis, but this mode can shift to lysogeny where the phage DNA gets integrated with host DNA and stays there as prophage till again it becomes lytic. 'How are these two states interconnected, and how does one state shift to another' are the subject of regulation of gene expression. A map of lambda (λ) DNA is given in Figure, which has early genes in the middle and late genes at both ends of linear DNA. The linear DNA takes a circular shape during infection, so that the late genes from both ends come together in a cluster to form one transcription unit. When the phage infects a host cell, the early and delayed early genes of the phage are expressed irrespective of whether the phage has to follow lytic or lysogenic mode of development. At this stage, a decision about alternative pathways of development is taken. If the late genes are expressed at this stage, lytic cycle sets in, but if a regulator gene *(cI)* synthesizing a repressor is expressed, the late genes cannot be expressed and lysogeny sets in (Fig.). It is interesting to note that repressor protein coded by *cI* regulates its own synthesis also by working as an activator, so that if repressor is inactivated, its further synthesis does not take place and the phage is forçed to enter a lytic cycle. Mutants in the gene *cI* cannot maintain lysogeny, since then the late genes cannot be stopped from being expressed, due to the absence of active repressor.

Three Sets of Genes in Phage Lambda λ

Immediate Early cro and N Genes and their Control Regions (P_R/O_R and P_L/O_L) : As soon as the phage infects, transcription is initiated with the help of two promoters, P_R on the right side and P_L on the left side. The transcription starts on both DNA strands, one strand being transcribed on the left side and the other on the right side. There are two immediate early genes cro and N, cro being transcribed on the right side under the control of P_R and N being transcribed on the left side under the control of P_L. When transcribed, cro has dual function, namely (i) it prevents synthesis of repressor by *cI* (an action which is necessary for lytic cycle), which otherwise causes

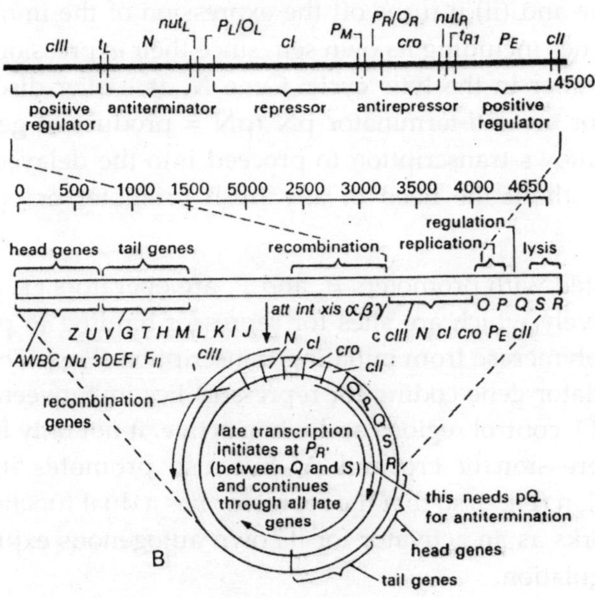

A map of lambda (λ) phage showing clustering of genes (A) and the circularization of DNA during infection (B) thus bringing together the genes on extreme left and extreme right.

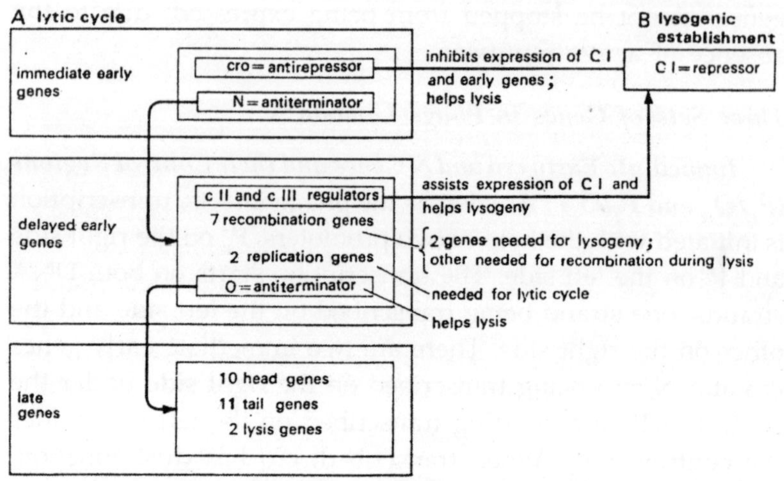

Regulation of the activity of three classes of genes in lytic cycle and in lysogenic establishment in lambda phage (for details see text).

lysogeny and (ii) it turns off the expression of the immediate early genes including its own self, since their expression is not needed later in the lytic cycle. Gene *N*, as earlier discussed, codes for an anti-terminator pN (pN = product of gene *N*), which allows transcription to proceed into the delayed early genes without the need of any fresh promoter or operator genes.

Associated with promoters P_R and P_L are operators O_R and O_L respectively which are sites for repressor binding to prevent RNA polymerase from initiating transcription (Fig.). The gene *cI* (regulator gene coding for repressor) lies in between P_R/O_R and P_L/O_L control regions and when active, it not only inhibits the expression of cro and A′ but also promotes its own expression (Fig.), so that the repressor has a dual function and also works as an activator for its own autogenous expression and regulation.

Delayed Early Genes (cII/cIII and Q) and their Control on Lysogeny and Lytic Cycle : Among the delayed early genes, regulators cII (on the right) and cIII (on the left) regulate synthesis of repressor by *cI*, to allow phage enter lysogeny (Figs.). On the other hand, a gene Q (on the right) is a regulator which acts as an anti-terminator, and allows the transcription of late genes meant for lysis. The two pathways (lytic and lysogenic) are 'so intimately related that it is difficult to predict which pathway will be followed and how is the decision for two alternative pathways is actually taken. This will be further discussed later in this chapter.

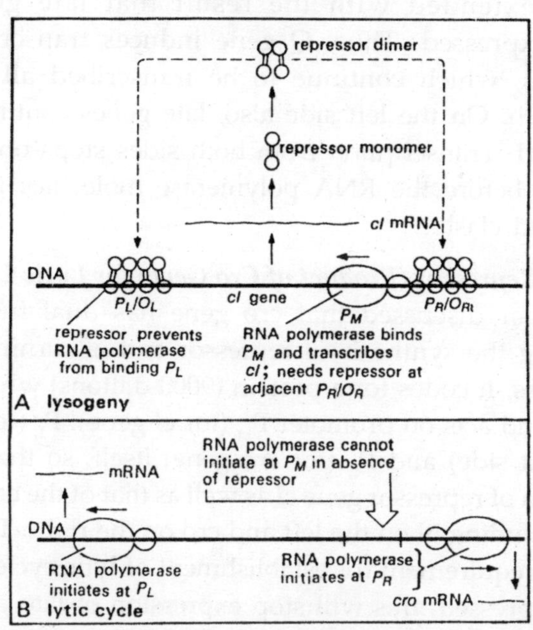

The maintenance of lysogeny by an autogenous circuit of repressor production (A), and the onset of lytic cycle due to interruption in this circuit caused by absence of repressor (B).

Late Genes for Lysis (Right End) and, Tail and Head Genes (Left End) : In the phage particle, the late genes are found at the two ends of the linear DNA molecule, genes S and R on the right end, and the tail and head genes on the left end. Genes for recombination are also found on the left side of *cIII* and are transcribed as late genes. Soon after the infection, the two ends of DNA molecule join to form a ring, so that all late genes are arranged in a single group containing S-R genes from right end and head and tail genes A to J from left end (Fig.).

There is a promoter gene P_R between genes Q and S. In the absence of pQ (product of Q gene), the transcription from P_R is constitutive, but terminates at t_{R3} lying close to P_R giving a product, 6S RNA (194 bases long).

However, when pQ is present, it suppresses t_{R3} and 6S RNA is extended with the result that late genes are heavily expressed. Thus, Q gene induces transcription of late genes, which continue to be transcribed all through their length. On the left side also, late genes continue to be transcribed. Transcription from both sides stops on the ring molecule, before the RNA polymerase molecules from two sides could clash.

Anti-Repressor (Product of Cro Gene) for Lytic Cycle : We have earlier discussed that cro gene has dual function of preventing the synthesis of repressor and of turning off the early genes. It codes for a protein (9000 daltons) which forms a dimer and acts on promoter P_M (for cI gene), P_L (for N gene on the left side) and P_R (for cro gene) itself, so that it stops expression of repressor gene *cI* as well as that of the early genes (Figs.) including N on the left and cro on the right. This fulfils the early requirement for establishment of lytic cycle, because if *cI* is expressed, this will stop expression of late genes and thus will establish lysogeny.

Lysogeny Through Foundation and Maintenance of Repressor (Promoters P_E and P_M) : In phage lambda (λ), a very delicate balance is maintained between lytic cycle and lysogeny, with a very sophisticated and intricate circuit of events. It will be seen that the repressor's synthesis is first established with the help of a promoter P_f under the control of genes *cII* and *cIII*, and is subsequently maintained by the repressor itself through a promoter P_M.

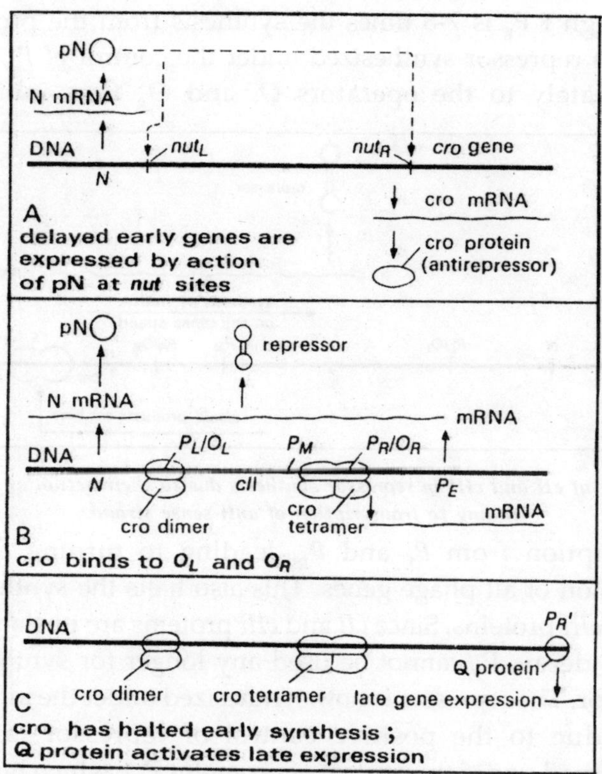

Synthesis and role of Cro protein in prevention of repressor maintenance both through PM (directly due to binding at OL and OR) and PE (indirectly due to switching off the delayed early genes), thus leading to lysis.

Promoter P_E (or P_{RE}) for Repressor Establishment and the Clear Plaque Genes (cI, cII, cIII) : There is a promoter gene P_E lying between *cro* and *cII* genes. The product of genes *cII* and *cIII* act on the promoter P_E, so that the RNA j polymerase enzyme can initiate transcription towards the left thus synthesizing RNA on genes j *cro* and *cI*. This transcript will consist of (i) the transcript from the anti-sense strand of cro gene, which cannot be translated (cro gene is transcribed in the right direction from P_R to yield translatable mRNA) and (ii) mRNA from *cI* gene which can be translated to give rise to repressor (Fig.). The efficiency of synthesis of repressor from *cI* through f P_E is 7-8 times the synthesis from the promoter P_M. This repressor synthesized under the control of $P_{E'}$ binds immediately to the operators O_L and O_R thus inhibiting

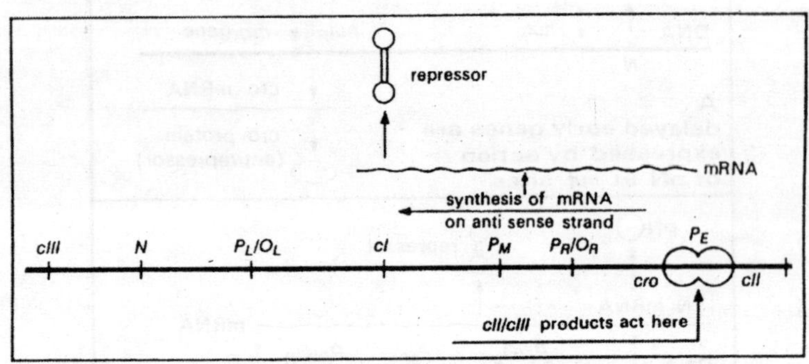

The role of cII and cIII in repressor synthesis due to their action at P_E site leading to transcription of anti-sense strand.

transcription from P_L and $P_{R'}$ leading to turning off the expression of all phage genes. This also halts the synthesis of *cII* and *cIII* proteins. Since *cII* and *cIII* proteins are unstable and rapidly decay, P_E cannot be used any longer for synthesis of repressor. The repressor is now synthesized under the influence of P_M due to the positive control of repressor (already synthesized under the control of promoter P_E), which binds on O_R (which also contains the promoter P_R).

The genes *cI, cII* and *cIII* are so named, because mutants in these genes produce clear plaques, due to the presence of only lysed cells. In the wild type phage, when the lysis occurs, due to the presence of some lysogenic bacteria, clear plaques are not obtained (plaque is a region on the culture medium on the plate, generated due to lysis of bacterial cells). Mutants in *cI* (*cI*⁻) differ from the mutants *cII*⁻ and *cIII*⁻, since the former can neither establish, nor maintain lysogeny. The mutants *cII*⁻ and *cIII*⁻ have difficulty in establishing lysogeny, but once established, lysogeny can be maintained. This suggested that the products of genes *cII* and *cIII* are needed only for the establishment of lysogeny, but not for its maintenance. The genes *cII* and *cIII* are thus needed to circumvent the difficulty of autogenous circuit of *cI*, which cannot start the synthesis of repressor de novo. The different steps involved in establishment and autogenous maintenance of lysogeny are shown in Figure.

Promoter P_M (P_{RM}) for Autogenous Maintenance of Repressor: Once the synthesis of repressor is established through the action of *cII* and *cIII* gene products on P_E, these genes (*cII* and *cIII*) are switched off by the action of repressor on O_R and O_L. But the repressor works as an activator by binding on O_R, which also contains the promoter P_M (also known P_{RM} since it

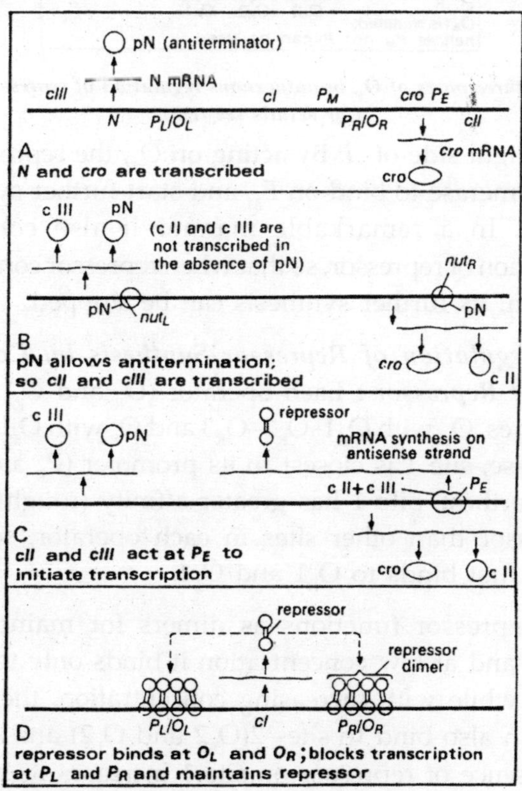

Establishment of lysogeny through a cascade involving the anti-terminator protein pN and its further maintenance by the onset of an autogenous repressor maintenance circuit.

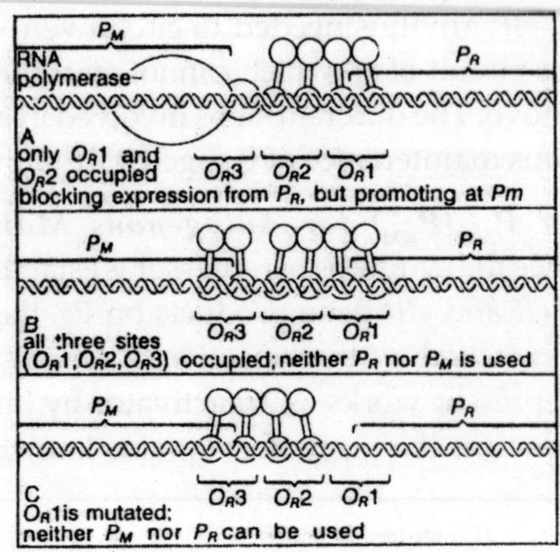

The role of three parts of O_R in autogenous regulation of repressor synthesis (for details see text).

is on the right side of *cI*. By acting on O_R, the repressor helps RNA polymerase to bind on P_M and start further synthesis of repressor. In a remarkable manner it also controls the concentration of repressor, so that when repressor concentration is too high, its further synthesis can be stopped.

Autoregulation of Repressor Synthesis by Cooperative Binding of Repressor : Each operator (O_R and O_L) has three binding sites, O_R with O_R1–O_R2–O_R3 and O_L with O_L1–O_L2–O_L3. In each case, site 1 is closest to its promoter (P_R and P_L) and site 3 is farthest. Site 1 has greater affinity (roughly tenfold) for repressor than other sites in each operator, so that the repressor first binds to O_R1 and O_L1.

The repressor functions as dimers for maintenance of lysogeny, and at low concentration it binds only to sites O_R1 and O_L1, while with increasing concentration, the repressor dimers can also bind to sites 2(O_R2 and O_L2) and 3(O_R3 and O_L3). Presence of repressor on site 1 increases its affinity for site 2 but this interaction does not proceed to site 3. When lysogeny is established, only sites 1 and 2 are occupied but site 3 is not occupied. Since P_R lies within O_R1 and P_L lies within

O_L1, occupancy of O_R1–O_R2 and O_L1–O_L2 physically blocks P_R and P_L respectively and inhibit the expression of all genes thus establishing lysogeny.

lysogenic repression dimers of repressor bind to DNA	carboxy terminal domains form dimer connector DNA aminoterminal domains bind operator
induction dimeric carboxy terminal domains released from DNA -bound amino terminal domains	cleavage by protease DNA
lytic cycle starts individual amino terminal domains dissociate from DNA	DNA amino terminal domains

The role of C-terminal domains in controlling the binding of repressor dimers to the operator site (through the N-terminal domain of repressor). (Note that due to cleavage of C-terminal domains, the N-terminal domains lose their affinity for binding).

The relationship of O_R and P_M is such that P_M (site for RNA polymerase binding) lies rather close to O_R2. When dimers are bound to O_L1–O_L2, the, dimer at O_R2 interacts with RNA polymerase in such a way that the latter can utilize P_M for transcription (Fig.). It is interesting that a repressor dimer bound on O_R2 can inhibit transcription from P_R, but promote transcription from P_M.

However, when repressor concentration increases to such an extent, that it may now occupy even O_R3 in addition to O_R1–O_R2, the transcription from *cI* is prevented. This effect depends on the fact that P_M lies within O_R3 and when O_R3 is occupied by repressor, it is not available for RNA polymerase.. When repressor synthesis stops due to occupancy of O_R3 by repressor, the concentration of repressor consequently falls, and O_R3 again becomes available for RNA polymerase and synthesis of repressor starts once again due to the presence of repressor on

O_R2. Thus repressor is an autogenous regulator of its own synthesis causing the inhibition by negative control at its high concentration and promoting by positive control at its low concentration.

Change from Lysogeny to Lytic Cycle by Cleavage of Repressor Dimers in the Connector Region : The repressor functions as a dimer and each unit (27,000 daltons) has (i) a N-terminal domain (1-92 residues), which binds to operator; (ii) a C-terminal end (133-132 residues), which enhances the affinity of N-terminal domain for binding and (iii) a connector region (93-132 residues).

The induction of a lysogenic prophage to enter the lytic cycle is caused by cleavage of repressor in the connector region between 111 and 112 amino acid residues. The cleavage releases the C-terminal domains, so that N-terminal domains now do not have sufficient affinity to remain attached to the operator. N-terminal domains thus dissociate, leading to lytic infection, because now the genes repressed earlier due to repressor, will be expressed causing lysis (Fig.). This transition from lysogeny to lysis, involves the expression of Q gene with the help of promoter P_R' giving its product pQ regulator which helps in the expression of late genes S and R (Fig.).

Binding of DNA of Cro and λ Repressor Proteins : The structure of lac repressor and CAP proteins, with respect to its DNA binding properties. A similarity of CAP with Cro and λ repressor proteins, discussed in this chapter, has also been demonstrated (Table). Cro consists of a single domain with three a helices (α, α_2, α_3) and $\alpha\beta$ sheet. It has a helix-turn-helix motif that binds the operator symmetrically (Fig.). λ repressor resembles Cro, but consists of two distinct domains and a connector (as discussed earlier). The N-terminal domain of λ repressor is shown in Figure.

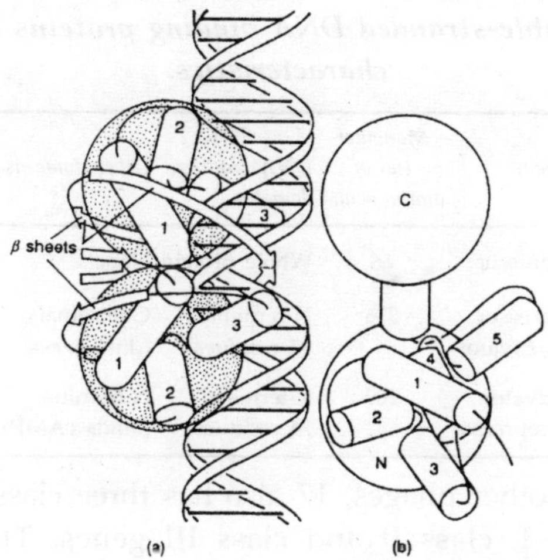

(a) Binding of Cro protein with operator (note that Cro consists of a single domain and its active form is a dimer; each monomer consists of three a helices and a β sheet); (b) N-terminal domain of λ repressor; α helices 2 and 3 are involved in binding DNA.

A helix-turn-helix model for DNA binding holds good in this case also. The amino terminal ends form two long arms, which wrap around the DNA (unlike Cro).

Grouping Function of Genes in Phages T_4 and T_7 : The bacteriophage T_4 has a large genome in which there are (i) essential genes and (ii) non-essential genes. These genes exhibit extensive functional grouping, but we do not fully understand the significance of the presence of so many non-essential genes.

There are three phases of gene expression : (i) early genes, (ii) quasi-late genes and (iii) late genes. The early genes and quasi-late genes are concerned with the synthesis of DNA, modifying cell structure or with transcribing and translating phage genes. Two of these genes involved in transcription are necessary for late gene expression. However, we do not know, how the cascade of three phases of gene expression is controlled.

Three double-stranded DNA binding proteins and their characteristics.

Protein	*Action*	*Monomer (no of amino acids)*	*DNA binding domains*	*Other domains*	*Actoce from*	*Size of DNA site (op)*
λ Cro	Repressor	66	Whole protein	None	Dimer	17
λ repressor (or CI)	Repressor or activator	236	N-terminal 92 residues	C-terminal (dimerizes)	Dimer	17
CAP	Activator or repressor	209	C-terminal 74 residues	N-terminal (binds cAMP)	Dimer	15

Like all other phages, T7 also has three classes of genes called class I, class II and class III genes. The enzymes synthesized on phage genes interfere with host gene expression and the synthesis of phage RNA polymerase. The phage enzymes synthesized by T7 genes are responsible for expression of class II genes and class III genes. The basis for the difference in expression between class II and class III genes is not known.

2

Operon Circuits

Different genes in an organism are meant for the synthesis of different proteins. All these proteins are not needed at one time. Further, the same proteins are not needed in different tissues. Specific enzymes are needed at different times and in different tissues in the life cycle of an organism. However, at all times in the life cycle, and in different tissues/organs, every cell contains same set of genes. It would be necessary, therefore, to have mechanisms which would allow only the desired genes to function at a particular time.

The activity of other genes will have to be restricted. This will require temporal (in time) as well as spatial (in space) regulation of the expression of individual genes. A variety of mechanisms are now known, which regulate gene expression at different levels including transcription, processing of mRNA and translation. Although information on regulation was available from bacteria for quite some time, only in recent years, work in this area became possible in eukaryotes.

Process of Induction and Repression

In *E. coli,* synthesis of P galactosidase, an enzyme meant for hydrolysis of lactose into glucose and galactose, has been studied in considerable detail, p galactosidase

$$\text{lactose} \xrightarrow{\beta\text{galactosidase}} \text{glucose} + \text{galactose}$$

If β galactosides (*e.g.* lactose) are not supplied to E. coli cells, the presence of β galactosidase is hardly detectable, but as soon as lactose is added, production of enzyme β galactosidase increases (Fig.) as much as 10,000 times. The enzyme quantity again falls down as quickly as the substrate

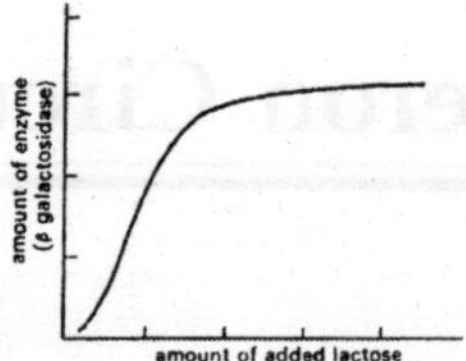

Induction of β galactosidase synthesis by lactose.

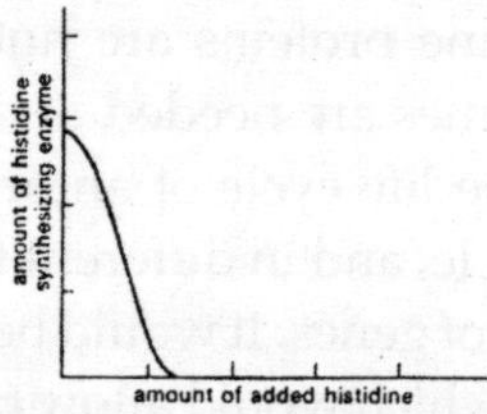

Repression of histidine synthesis enzymes by histidine.

(i.e. lactose) is removed. Such enzymes, whose synthesis can be induced by adding the substrate are known as inducible enzymes and the genetic systems responsible for the synthesis of such an enzyme are known as inducible systems.

In some other cases the situation is reverse. For instance, when no amino acids are supplied from outside, E. coli cells can synthesize all the enzymes needed for synthesis of different amino acids. However, if a particular amino acid like histidine is added, production of histidine synthesizing enzymes will fall down. In such a system the addition of end product of a bio-synthetic pathway will check synthesis of the enzymes needed for its biosynthesis. Such enzymes whose synthesis can be checked by adding end product are known as repressible enzymes and these genetic systems would be known as repressible systems.

Role of Inducer and Co-repressor

The substrate whose addition induces synthesis of an enzyme, (as lactose in case of the synthesis of β-galactosidase), would be called inducer. Similarly, the end product, whose addition will check the synthesis of bio-synthetic enzymes (as histidine in case of histidine bio-synthetic enzymes), is known as co-repressor.

Although ordinarily only a substrate acts as an inducer, but rarely there are molecules that adequately resemble the natural inducers, but are not metabolized themselves by the enzyme. An excellent example is isopropyl thiogalactoside (IPTG), which, like several other galactosides, resembles lactose and thus has the property of induction. However, IPTG is not recognized by β-galactosidase. Such inducers which induce enzyme synthesis, without getting metabolized are called gratuitous inducers. Such inducers are very useful for conducting recombinant DNA studies, since they induce enzyme synthesis, but remain in the cell in original form. This also indicates that the substrate should have sites other than those for target enzyme, which will help the substrate in inducing enzyme synthesis.

We learnt above that in the absence of lactose no β-galactosidase is synthesized. This would mean that in the absence of 'inducer' the gene or genes responsible for the synthesis of β-galactosidase do not function. How is it brought about ? It is proposed and verified that a class of molecules called repressors are found in cells and these repressors check the activity of genes. Ordinarily, these repressors may be active or inactive. Active repressor may be rendered inactive by addition of an inducer, while an inactive repressor can be made active by addition of a co-repressor.

(1) *Inducible system :*

active repressor + inducer = inactive repressor.

(2) *Repressible system :*

inactive repressor + co-repressor = active repressor.

Operon Style for Regulating Transcription

F. Jacob and J. Monod in 1961, on the basis of their study on the inducible system for the synthesis of β-galactosidase enzyme in E. coli, proposed a model in order to explain the induction or repression of enzyme synthesis. The model, popularly known as operon model, has been variously modified, ever since it was originally proposed by Jacob and Monod. In its original form 'operon model' assumed the presence of a unit of coordinated control of protein synthesis and this unit was called 'operon'. An operon consisted of (i) an operator gene which controlled the activity of (ii) a number of structural genes which take part in the synthesis of protein(s). This means that the structural genes will synthesize mRNA under the operational control of an operator gene. The operator, in its turn, is under the control of a repressor molecule synthesized

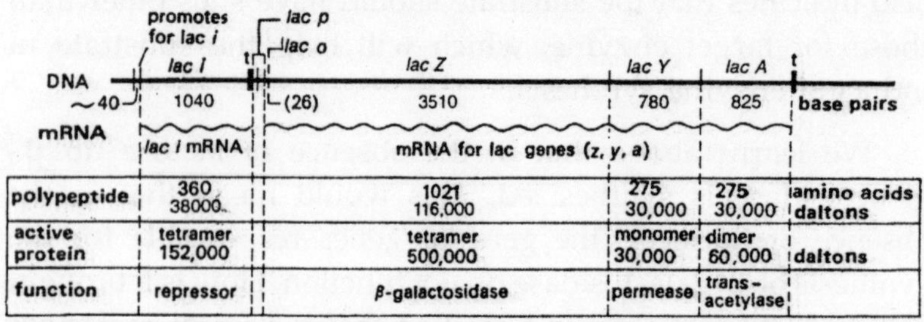

The lactose operon of E. coli and its regulatory gene (redrawn from Lewin's Genes IV).

by a regulator gene, which is not a part of the operon. The mechanism of regulation of protein synthesis utilizing operon model can be illustrated using the example of two different operons (lac and tryptophan operons) in bacteria.

The Lac Operon in E. coli : The lac operator consists of 26 base pairs and is a part of the first structural gene lacZ (meant for β galactosidase). There are three structural genes in lac operon which includes the following : (i) gene lacZ (3510 bp) codes for the enzyme β galactosidase, which is active as a tetramer and breaks lactose into glucose and galactose to be utilized in the cell; (ii) gene lacY (780 bp) codes for β galactose permease, which is a membrane-bound protein and helps in transport of metabolites and (iii) gene lacA (825 bp) which codes for β galactose transacetylase, an enzyme that transfers an acetyl group from acetyl-CoA to β galactosides. Any mutations in lacZ and lacY genes give lac- genotypes, which cannot utilize lactose. While lacTT mutants abolish enzyme activity, lacY⁻ mutants can not take up lactose from the medium. No defect is observed in lacA⁻ mutants, which is a puzzle. It is possible that acetylation gives an advantage for utilizing certain non-metabolizable analogues of β -galactoside, because it results in detoxification and excretion.

In 'lac' operon of E. coli, repressor is synthesized through activity of a gene lacl, known as regulator gene. The repressor then attaches on operator gene and checks synthesis of mRNA. If lactose is added, the repressor is rendered inactive, so that it cannot attach on operator gene, and the synthesis of mRNA takes place (Fig.). The activity of the regulator gene is controlled

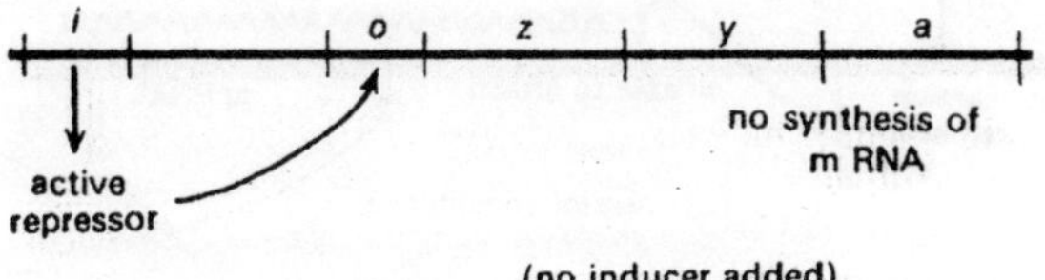

A 'lac' operon showing the action of repressor.

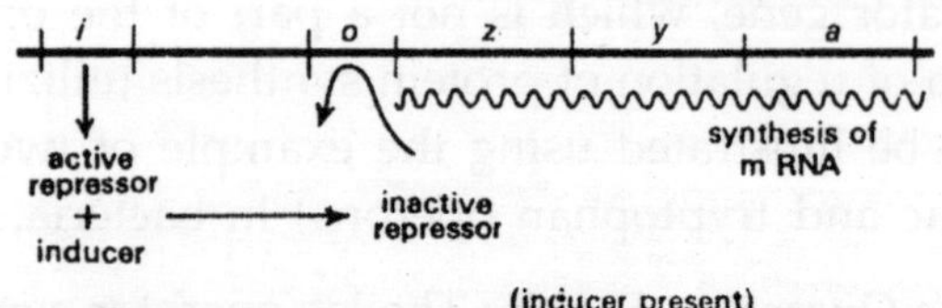

A 'lac' operon showing induction by inducer.

by a promoter gene on its left side. It has been shown that mutations in this promoter may increase the quantity of mRNA for repressor. This promoter is different from the promoter for lac operon which is described later in this chapter.

Lactose is not the Real Inducer of Lac Operon : Since in the repressed state of lac operon, E. coli cells will not synthesize β galactoside permease, one can question how lactose enters the cell. It is believed that a minimal amount (basal level) of enzyme is always present in the cell to start the metabolic pathway. It has been shown that lactose is not the real inducer of lac operon, but rather it is an anti-inducer, since it binds to repressor to increase its affinity for operator. On the other hand, the bound protein in inactive repressor was allolactose. It has also been shown that while β galactosidase breaks lactose into glucose and galactose, a side reaction changes galactose into allolactose and glactobiose. Therefore, in the

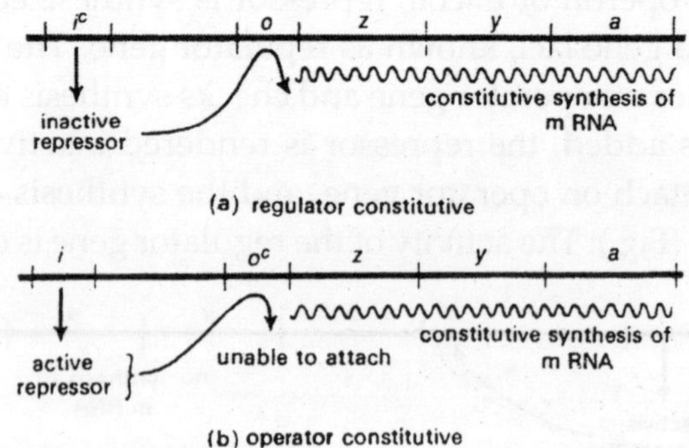

Constitutive strains resulting due to mutations in regulator gene and operator gene (redrawn from Watson : Molecular Biology of the Gene).

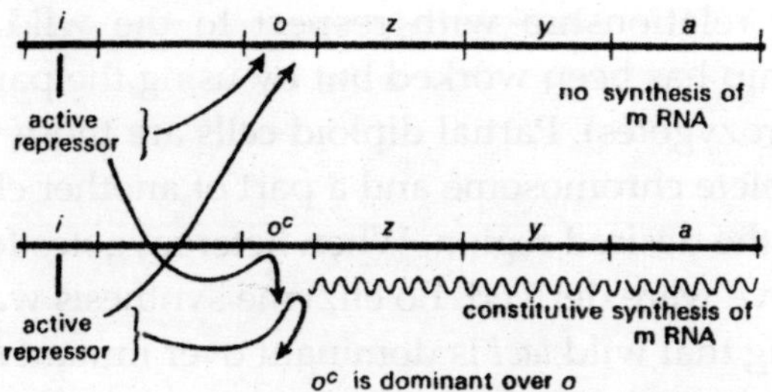

Check on enzyme synthesis in a partial diploid, heterozygous for regulator gene (I/I^c) (redrawn from Watson : Molecular Biology of the Gene).

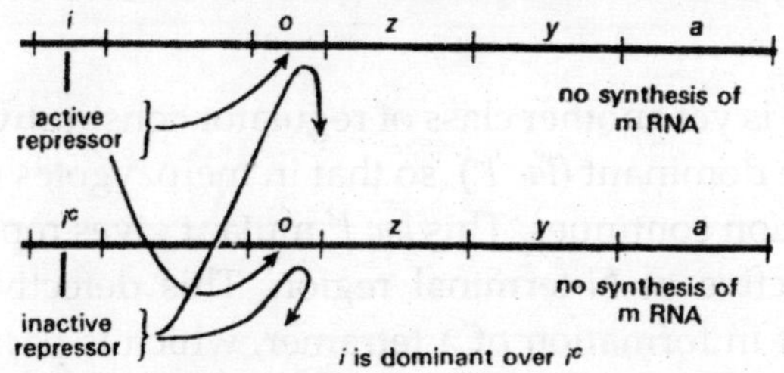

Constitutive enzyme synthesis in a partial diploid, heterozygous for operator gene ($0/0^c$) (redrawn from Watson : Molecular Biology of the Gene).

regulation of lac operon, lactose is taken up and a part of it, through galactose, is converted into allolactose, which then induces the operon, and outweighs the anti-inducing effect of lactose.

Constitutive Mutants : It has been found that a mutation in regulator gene lacl results in the absence of active repressor and consequently in uncontrolled synthesis of enzyme. Such strains which continuously synthesize enzyme independent of need are known as constitutive strains or constitutive mutants. These constitutive strains can also be obtained by mutations in the operator gene. A mutation in operator gene does not allow repressor to act because the repressor functions by acting on operator site. Therefore, there are two types of constitutive strains : (i) regulator constitutive and (ii) operator constitutive.

The above two types of mutations, though have the same net result, differ in their location as well as in their dominant-recessive relationship with respect to the wild type. This relationship has been worked but by using the partial diploid cells (merozygotes). Partial diploid cells are those which have one complete chromosome and a part of another chromosome covering the desired region. When heterozygotes for regulator constitutive were derived, no enzyme synthesis was observed suggesting that wild *lacI* is dominant over mutant *lacI*^{c}. On the other hand, partial diploid cells heterozygous for mutation at operator site exhibited constitutive synthesis of enzyme, indicating that normal operator gene 'O' is recessive to mutant gene *'0*^{c}*'*.

There is yet another class of regulator constitutive mutants, which are dominant (*lac I*^{d}), so that in merozygotes (*lac I/lac I*^{d}), transcription continues. This *lac I*^{d} mutant gives repressor sub-unit defective at N-terminal region. This defective sub-unit takes part in formation of a tetramer, which is partly derived from *lacI* and partly from *lacI*^{d} , so that the tetramer (the form in which repressor binds) is unable to bind on the operator.

Promoter Gene : While operator gene is the site of action of repressor, the actual site of transcription activation is different and is known as promoter gene, which lies to the left of operator gene. It is also believed, therefore, that perhaps RNA

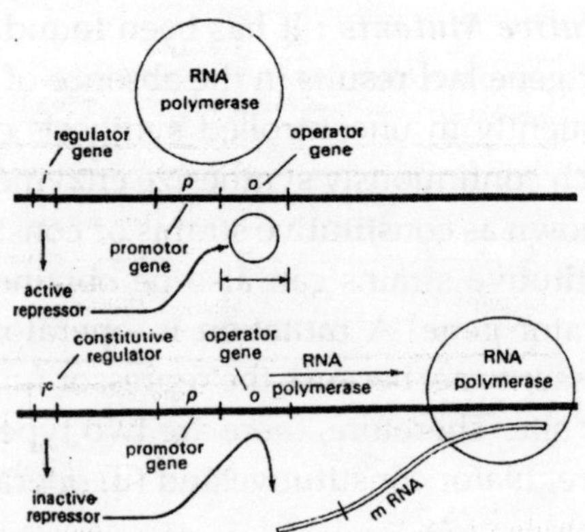

Relationship between promoter gene, repressor and RNA polymerase (redrawn from Watson : Molecular Biology of the Gene).

polymerase binds to and moves from promoter site. The repressor, when attached to operator, prevents the movement of RNA polymerase. The promoter gene in lac operon was inferred from mutations at this locus, which were found to affect the maximum rate of transcription.

Thus, there are two proteins which are known to be involved in the regulation of lactose operon : (i) lac repressor and (ii) catabolite gene activator (CGA) protein, also known as CAP (cyclic AMP protein), or CRP (cyclic AMP receptor protein). While repressor binds to operator site, cga protein binds to cga site. Moreover, repressor exercises a control by checking RNA polymerase to travel through operator, but cga protein exercises a positive control. Only when cyclic AMP molecule activates cga protein, the latter will allow RNA polymerase to bind (Fig.). A protein such as CAP, which interacts with DNA and RNA polymerase to assist in transcription initiation is called apoinducer.

It has been shown that cyclic AMP is essential for activation of cga protein, because when cyclic AMP was removed or was absent due to mutation, transcription was reduced. In mutants lacking cyclic AMP, transcription was restored by adding cyclic AMP exogenously. It has also been shown that this cyclic-AMP—cga protein system is influenced by the level of glucose concentration. It is, therefore, sometimes described as glucose sensitive system which is common to several glucose sensitive operons in E. coli (e.g. operons for degradation of lactose, galactose, arabinose, maltose, etc.). In all these cases, glucose is one of the catabolic products of bio-synthetic pathway and if accumulates, it is injurious. Therefore, whenever the concentration of glucose increases beyond a certain level, it leads to reduction of transcription. This is achieved by reducing the level of cyclic AMP by glucose, either through checking the synthesis of adenocyclase, an enzyme needed for synthesis of cyclic AMP or by removing cyclic AMP, by making a complex with it. Therefore, transcription in lac operon requires that (i) cyclic AMP activates cga protein—a positive control and that (ii) lac repressor is inactivated by inducer—a negative control.

Operator and Promoter are Cis-dominant : It should be realized that in the lac operon, operator can control only lac genes that are adjacent to it. If a second lac operon is introduced as done in partial diploids or merozygotes, none of the operators is influenced by the other operator. For instance, if in a partial diploid, one operon has normal operator (*O*) and the other has a mutant operator (O^C), the operon with O^C will be expressed, but that with *O* will not be expressed. This could be tested by the using partial diploids which not only differ in operator, but also differ in structural genes. For instance, in a partial diploid with genotype $O^CZ^+Y^-/OZ^-Y^+$, only β galactosidase will be synthesized, but not the (i galactoside permease, although Y^+ is present along with O^c in the same cell, but on two different chromosomes. This phenomenon of lack of dominance extended to other chromosome (trans configuration) is called cis-dominance. This also suggests that operator does not act through a gene product like regulator gene which acts through repressor, that can diffuse and influence the other allele also. The phenomenon of cis-dominance will be expressed by all those sites, which are physically contiguous with the sequences of controls, and function by being recognized rather than by giving a gene product. It is for this reason that promoter and terminator are also cis-dominant or cis-acting, while regulator is transacting.

Uninducible Mutants (Promoter and Regulator Genes): Uninducible mutants in the lac operon are those which cannot be induced into synthesis of enzyme even on lactose addition. These can be of two types, cis-acting and transacting. (i) Cis-acting mutations may be in the promoter site (see later), since they act by preventing RNA polymerase from binding to the

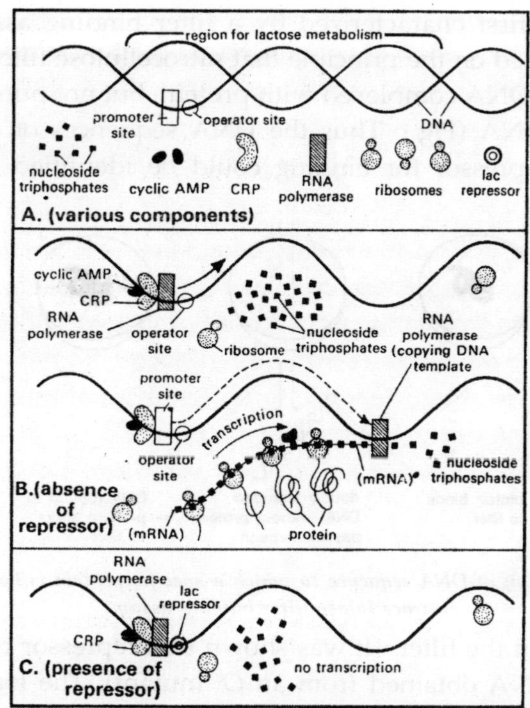

Mechanism involved in the positive control system for the regulation of gene activity in E. coli lac operon. Note that only in the absence of the repressor, RNA polymerase enzyme can travel and transcribe lac operon as shown in B. The repressor, when present on operator site is an obstacle in the path of RNA polymerase.

promoter, so that the operator can not function due to non-availability of RNA polymerase. (ii) Transacting uninducible mutants lie in the regulator gene, since they act by changing the repressor protein so that the altered repressor cannot bind the inducer (lactose), and will always be 'locked in' to the active form for attachment to the operator. These mutations are designated as '*lac* I^s' and are dominant over wild type. This happens because mutant repressor binds to all wild type operators in the cell to prevent their transcription.

Association and Dissociation of lac Repressor with Operator. The binding of purified lac repressor protein to

DNA was first characterized by a filter binding assay. The assay is based on the principle that nitrocellulose filter retains protein, or DNA complexed with protein, but not pure double stranded DNA (Fig.). Thus the DNA sequences of operator used by repressor for binding could be identified by their

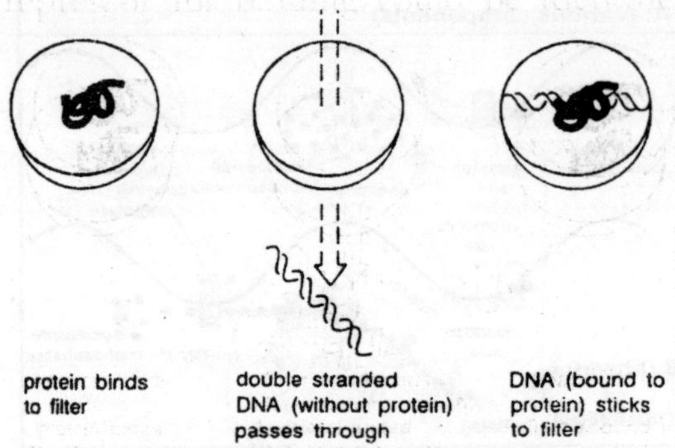

Identification of DNA sequence to which a specific protein is bound, by nitrocellulose filter binding assay.

retention on the filter. (It was shown that repressor could not bind on DNA obtained from an O^c mutant). The location of binding in lac operon extended from – 5 to + 21 within the transcription unit, thus overlapping the right end of promoter (this overlapping does not occur in all operons). Although earlier it was believed that binding of repressor does not allow binding of RNA polymerase, now we know that both can bind simultaneously, but RNA polymerase can not initiate transcription, unless repressor is released by an inducer.

Repressor is a polypeptide of 360 amino acids and is divided into two regions : (i) 1-59 amino acid residues on the amino terminal end make a long head piece, which can be cleaved at position 51, and the fragment 1–51 is called short head piece, (ii) 60-360 amino acid residues on the C-terminal end make trypsin resistant core (trypsin cleaves the repressor at amino acid no. 12). The long and short head pieces both can bind to operator independently of the core, and therefore, it is assumed that their structure does not change in association with the core. Several models have been proposed suggesting the manner of association of repressor, but all models assume that head pieces make protrusions, which help in binding.

When lactose inducer is added, repressor will dissociate from the operator. This is explained on the basis of several models : (i) Repressor is normally in equilibrium between associated and dissociated state. When the inducer binds to dissociated repressor, it is rendered incapable of association, (ii) Inducer may bind to repressor associated with operator and changes the conformation of repressor, which allows it to

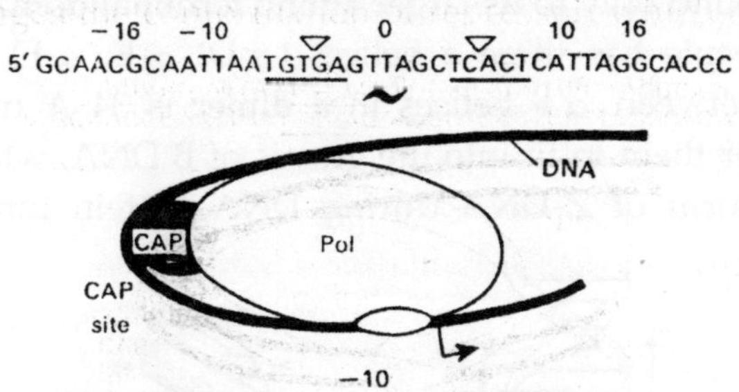

DNA bending due to binding of CAP protein.

get off the operator, (iii) Inducer changes the relative conformation of head and core regions of repressor, so that head piece cannot hold the DNA tightly, (iv) The repressor in tetrameric form may be in excess and remains associated at other random sites on DNA, but is not found free within the cell. Due to induction, when the repressor is dissociated from lac operator, it is stored at other random sites and used again when needed. The effect of induction, therefore, is to change the distribution of repressor on DNA and not to release free repressor.

DNA Bending During Binding of CAP : protein. During late 1980's and 1990's, DNA-protein complexes have been analysed to find if DNA geometry is altered, when a regulatory protein binds to it. It has been shown that DNA bends considerably, when the regulatory protein CAP binds to promoter site. Similar bending has been observed during binding of zinc finger proteins or leucine zipper proteins or TFIID. This bending of DNA is very important in activation of transcription. It has been shown that bending due to CAP alone can be about 90°, but the protein-protein interaction between CAP and RNA polymerase leads to formation of a nucleoprotein complex showing a DNA bending of upto 180° (Fig.).

The CAP protein consists of 209 amino acid residues with two domains. The carboxyl terminal domain of the protein, which is involved in DNA binding, is shown in Figure. Cyclic AMP is bound only to its larger amino terminal domain. Each CAP molecule has three a helices (α D, α E, α F) and the distance between α F helices in a dimer is 34 A making it possible for them to fit into the groves of B DNA, which may take the form of Z-DNA during DNA-protein interaction.

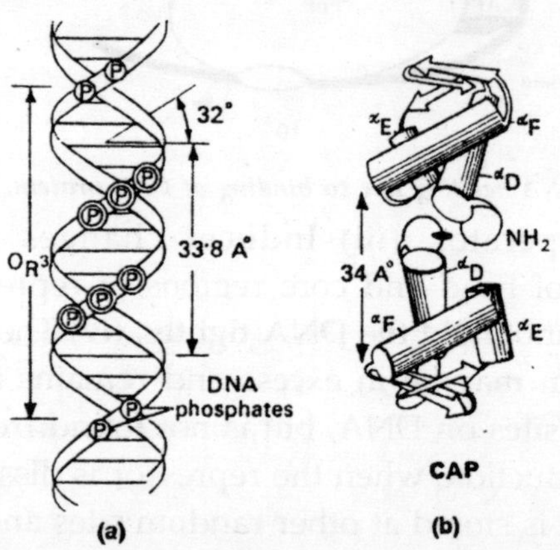

A DNA segment and its relationship (for binding site) with CAP protein involved in regulation of gene activity. Note that CAP consists of a dimer, each monomer having three polypeptides.

CAP has been compared with other proteins like λ repressor and A, cro protein, showing similarities in structural details and amino acid sequences in DNA-binding domains. Comparisons of these three proteins (CAP, λ Rep and λ Cro) revealed a conserved recognition motif consisting of an

α -helix, a turn and second α -helix. These are therefore described as helix-turn-helix or HTH proteins.

CAP, λ Repressor vs. NTRC (Recruitment of RNAP vs. Interaction with pre-bound RNAP : As discussed earlier, CAP stimulates promoters for lac and gal operons and λ repressor stimulates promoter P_{RM} in λ phage. In contrast, NTRC is a regulatory protein (found in bacteria) that stimulates an enhancer sequence and requires ATP (e.g. gln Ap2 gene). Further the RNA polymerase bearing a is involved with CAP and λ repressor, while RNA polymerase bearing σ^{54} (used under nitrogen starvation conditions) is involved with NTRC. In E. coli, RNA polymerase bearing a binds stably to the promoter of glnAp2, without the help of an activator, while that with σ^{70} binds to lac promoter or A. promoter P_{RM} only with the help of an activator. It has been shown that while CAP and λ repressor recruit RNA polymerase to DNA, NTRC (and its relative NifA) works on a stably bound polymerase. As shown earlier, at most bacterial promoters, polymerase binds in at least two steps: in the first step, it binds to helical DNA to form a 'closed' complex and in the second step, the complex isomerizes to form the 'open' complex. The stability of the 'closed' complex and the requirements for the transition from 'closed' to the 'open' state in NTRC-stimulated complex differ from those in CAP and X repressor-stimulated complexes. It has also been shown, that if NTRC is available at sufficiently high concentrations, it will activate the target genes even without binding any DNA sequences.

The Tryptophan Operon in Bacteria (E. coli and Salmonella)

The five structural genes involved in the synthesis of tryptophan form an operon with an operator. These five genes are arranged in a series and code for three enzymes needed for tryptophan synthesis (Fig.). Transcription starts at the promoter site on the left of the structural genes. A repressor is coded by a gene trpR, which is not linked with the operon.

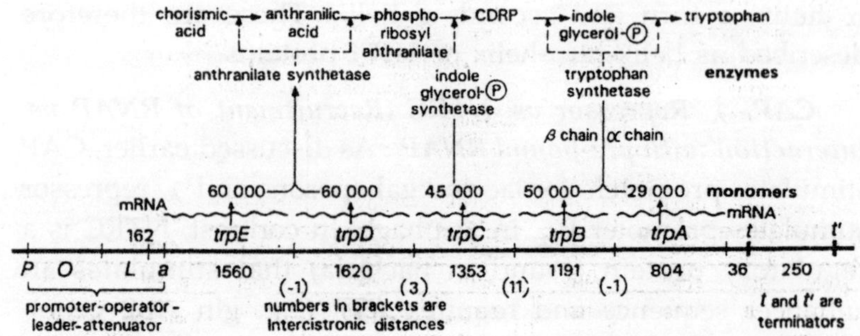

Tryptophan operon with its five genes, the promoter-operator leader-attenuator region and the corresponding bio-synthetic pathway. (CURP = carboxy phenylamino deoxyribulose phosphate).

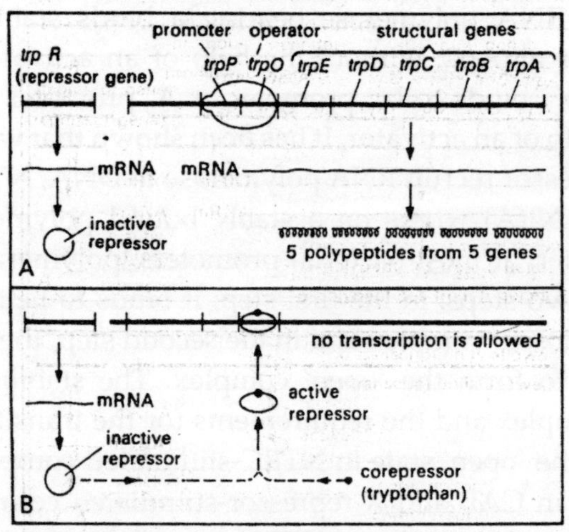

Tryptophan operon, showing transcription in the absence of corepressor (tryptophan) (as shown in A) and the prevention of transcription due to addition of corepressor, which activates the inactive repressor (as shown in B).

The repressor has low specificity for the operator and thus remains inactive, but can become active due to association with tryptophan which functions as a co-repressor (Fig.). The repressor functions as a tetramer of four identical sub-units of 12,000 daltons. The level of tryptophan, besides working as a co-repressor, also functions through feedback inhibition (see later in this chapter), where tryptophan combines with the first enzyme of tryptophan synthetic pathway and inhibits its activity through allosteric modifications (Fig.).

It is thus obvious that the synthesis of tryptophan is self regulatory, since tryptophan, when present in plenty, works firstly by feedback inhibition and secondly through co-repression to stop further synthesis of tryptophan. The major difference between lactose operon and tryptophan operon is that tryptophan increases the affinity of repressor protein for the operator. In tryptophan operon there is yet another mechanism for regulation of tryptophan synthesis through leader sequences and attenuation, which is discussed later in this chapter.

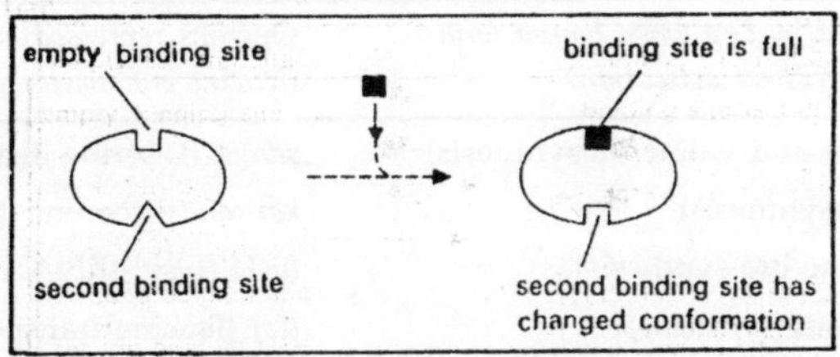

The change in conformation of the second binding site in an allosteric protein (monomer) when the first site is occupied.

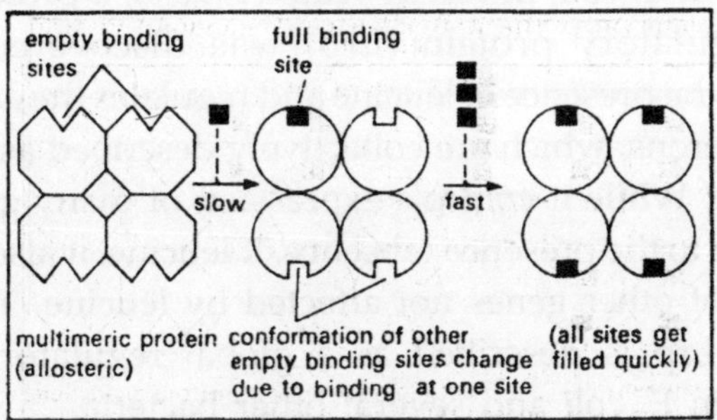

The change in conformation of all the empty sites in an allosteric protein, when one site is occupied by a slow reaction leading to quick occupation of these empty sites.

Tryptophan (trp) Repressor Controls Three Sets of Genes: The lac repressor acts only on lac operon with ZYA cluster of genes. However, a repressor can control several groups of genes dispersed in the genome. For instance, trp repressor controls following three sets of unlinked genes : (i) *trpEDBCA* meant for tryptophan synthesis enzymes; (ii) aroH gene coding for an enzyme needed for common pathway of aromatic amino acid biosynthesis, and (iii) *trpR* regulator gene (repressed by its own gene product = autogenous control). A related operator sequence (27bp long) is present at each of these three genes/ gene clusters, although the location of this sequence relative to start point may differ (as shown earlier). The operator may actually lie within the promoter (e.g. trpP locus), or downstream (e.g. he) or upstream (e.g. gal).

Some E. coli operons regulated by Lrp and affected by leucine.

Operons activated by Lrp when bound with leucine (leucine reduces activation)	*Operons repressed by free Lrp (leucine enhances repression)*
ilvIH (isoleucine and valine biosynthesis)	*sda A* (L-serine deaminase)
serA (serine biosynthesis)	*kbl-tdh* (threonine degradation)
leuABCD (leucine bio-synthesis)	*lys*U (lysyl-tRNA synthetase)
getBDF (glutamate synthase)	*livJ* (leucine transport)

Leucine Responsive Protein (Lrp) as a International Regulator : In E, coli, in recent years (1990's), a protein called leucine regulatory protein (Lrp) was discovered, which responds to the presence of leucine and regulates the expression of many operons, which are collectively described as leucine/ Lrp regulon. While it controls expression of many genes that are sensitive to the presence/absence of leucine, it also controls expression of other genes not affected by leucine. Therefore, sometimes Lrp is described as a global regulator of gene expression in E. coli and several other bacteria.

Lrp is a DNA-binding protein, which does not bind single-stranded DNA. It consists of two sub-units (each 1800 daltons) which make a dimer. The native Lrp protein has 163 residues and a sequence centred at 40 may represent helix-turn-helix (HTH) motif. In a cell, there are as many as 3000 molecules of this protein. This number is much higher than the number of any other regulatory protein like lac or A, repressor. This abundance indicates that Lrp may also function as a chromosome (DNA) organizer.

Lrp is a transcriptional regulator; when bound with leucine it functions as an activator, and when free (devoid of leucine) it functions as a repressor. Further, when it functions as an activator, its effect is reduced by leucine and when it functions as a repressor, its effect is enhanced by the presence of leucine. It is also known that some of the genes that are activated by Lrp are involved in bio-synthetic pathways, and some other genes that are repressed, are involved in degradation (Table). Lrp has been shown to affect at least 11 operons directly (e.g. gltBDF) and several others indirectly (e.g. glnA). It also regulates its own synthesis and therefore, is also an example of an autogenously regulated protein.

Negative and Positive Transcription Controls

The system of regulation in lactose and tryptophan operons outlined above is essentially a negative control in the sense that the operon is normally 'on but is kept 'off' by the regulator gene. In other words, the gene is not allowed to express unless required. However, it is shown that through cga protein (or CAP) and cyclic AMP, a positive control is also exercised in lac operon (Fig.). In a positive control the regulator gene will stimulate the production of enzyme as in case of arubinose operon of E. coli. The gene without the activator will be inactive in this case (Fig.). Most systems on regulation, studied initially, involved a negative control. The positive control systems were studied only later.

In arabinose operon (also known as OBAD operon) shown in the Figure, *O, B* and A are structural genes, *i* is the initiating site, O is the operator site and C is a gene comparable to regulator gene of lac' operon. Through the use of two mutations, C (which is characterized by reduction in transcription) and

C^c (which is characterized by the constitutive synthesis of enzymes), it could be shown that the gene C gives rise to a protein P,, which works as a repressor in the absence of

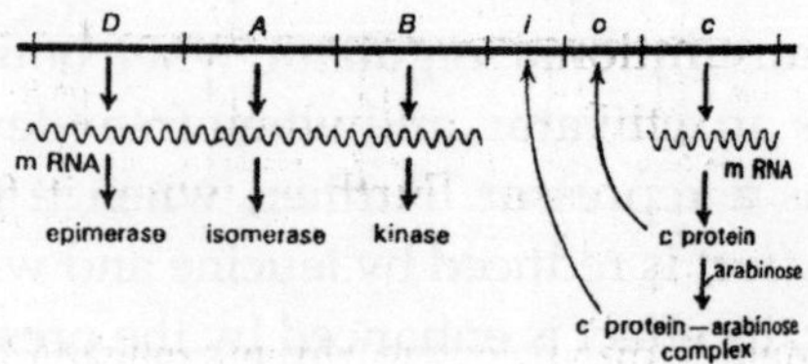

A map of arabinose operon showing positive control I of protein synthesis in E. coli.

arabinose. In the presence of arabinose, P, is converted into P_2, a protein working as a stimulator in transcription. The role of P1 as repressor was confirmed when it was observed that $araC^+$ is dominant over *ara*C^C, because if P_1 is not a repressor, then *ara*C^C should have been dominant over *ara*C^+. In this manner, arabinose operon, which was once thought to have only positive control, is now known to have both a positive and a negative control system.

Replacement of Sigma Factor and Transcriptional Control: We earlier discussed the role of sigma (a) factor in recognition of promoter site. Therefore, whenever more than one sigma factors are available for the same core enzyme, they can be used for recognition of different promoters. Substitution of sigma factor has been shown to result in switching 'on' certain genes in E. coli. In some other bacteria, it leads to change in the style of life from vegetative phase to sporulation phase. Substitution of sigma factors is also used for regulation of gene expression in bacteriophages during transition from lysogenic phase to lytic phase.

Multiple Sigma Factors in E. coli : In E. coli, no change from vegetative phase to sporulation phase occurs, and therefore for a long time it was believed that it has a single sigma factor, o . However, in recent years, new sigma factors in E. coli have been discovered, which operate under conditions of nutrient starvation (σ^{415} or σ^{5}), heat shock (σ^{32} or σ^{H}), nitrogen starvation (σ^{54} or σ^{N}) or for chemotaxis and flagellar structure (σ^{28} or σ^{F}). The details of these sigma factors are given in Table.

E. coli sigma factors, which recognize promoters with different consensus sequences.

Sigma factor	Gene	Mass (daltons)	Consensus sequences			Separation (bp)
			Use	– 35 sequence	– 10 sequence	
σ^{D} or σ^{70}	*rpo D*	70,000	general	TTGACA	TATAAT	16—18
σ^{S} or $\sigma^{41.5}$	*rpo S*	41,500	starvation	GTTAAGC	TATACT	—
σ^{H} or σ^{32}	*rpo H*	32,000	heat shock	CNCTTGAA	CCCCATNT	13—15
σ^{N} or σ^{54}	*rpo N*	54,000	nitrogen	CTGGNA	TTGCA	6
σ^{F} or σ^{28}	*flaI*	28,000	flagellar	TAAA	GCCGATAA	15

Multiple Sigma Factors in B. Subtilis : In bacteria like Bacillus subtilis, at the end of the vegetative phase, nutrients in the medium get depleted and a series of morphological changes are initiated. This involves segregation of a genome at one end of the cell, where eventually it gets surrounded by a tough spore coat. This process takes about eight hours and involves drastic changes in the bio-synthetic activities of the bacterium. Synthesis of RNA continues throughout sporulation and its inhibition brings the process to an end. During this process some of the genes functioning during vegetative phase are turned off, although most genes still function. But a large number of genes specific for sporulation are expressed only during this period. At the end of sporulation 40% of mRNA is specific to sporulation.

Genes specific for sporulation can be identified by spo-mutations, which will block sporulation. These mutants may be *spo0,* where sporulation does not start at all; others could be spoII, etc. where later stages are affected. These genes either encode enzymes or structural proteins or are involved in transition from vegetative phase to sporulation in some other manner (see below).

Some of the changes which occur during change of lifestyle from vegetative to sporulating phase are now known and have been studied to indicate that alterations in RNA polymerase are involved in the initiation of sporulation. In RNA polymerase of B. subtilis, at least three sigma factors namely σ^{43} (43,000 daltons), σ^{37} (37,000 daltons) and σ^{29} (29,000 daltons) have been identified, σ^{43} is involved in transcription during vegetative phase, while σ^{37} and σ^{39} are involved during sporulation. At the beginning of sporulation σ^{43} is replaced by σ^{37} (perhaps due to the activity of any of the five *spo0* genes), which either removes σ^{43} or modifies the core enzyme. The

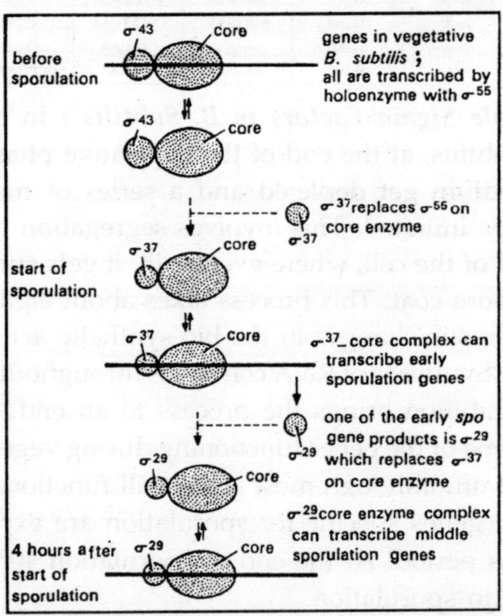

Changes in the initiation specificities of RNA polymerase by replacement of sigma factors during sporulat ion in Bacillus subtilis.

displaced σ^{43} can be recovered from sporulating cells. Therefore, $\alpha_2\beta\beta'\sigma^{37}$ initiates the transcription of gene needed for sporulation. About four hours after sporulation, another protein σ^{29} appears and RNA polymerase $\alpha_2\beta\beta'\sigma^{29}$ functions for transcription. Several other sigma factors have been isolated and characterized in B. subtilis. Thus, it is obvious that selective expression of genes can be caused by changes in sigma factor of RNA polymerase enzyme.

Controlling Transcription of DNA Sequences

Footprinting for Identification of DNA Sites used for Protein Binding : As pointed out earlier, the binding site on DNA for lac repressor was characterized by filter binding assay (Fig.). However, the technique of footprinting has been extensively used for identification of DNA sites used for binding specific proteins leading to regulation of gene expression. Following steps are followed using two samples of DNA to be

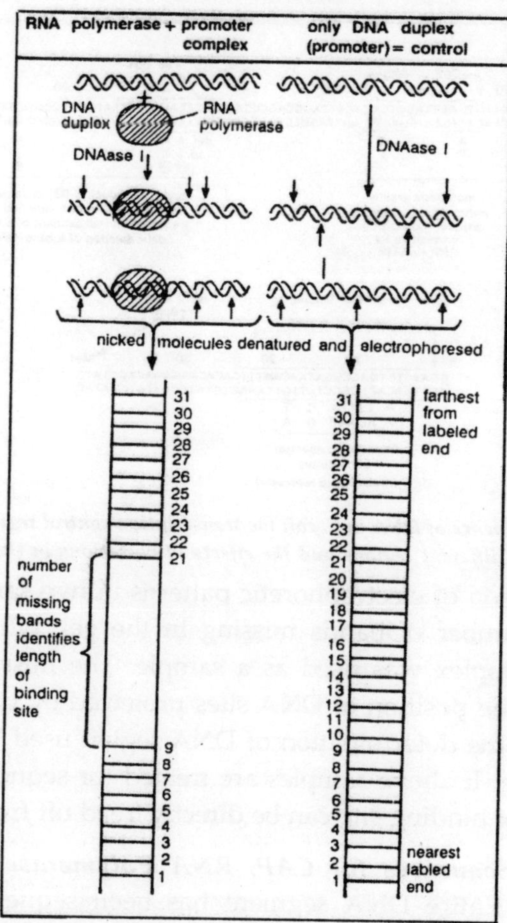

Identification of DNA binding sites for protein by the technique of DNA footprinting.

examined (one of them used as control and the other complexed with protein, whose DNA binding sites need to be determined): (i) DNA is end labelled; (ii) labelled DNA samples are subjected to partial cleavage (so that all cleavage sites are not attacked) with DNAaseI; (iii) population of nicked DNA molecules in both samples recovered by removing protein from the complex; (iv) DNA segments are denatured; (v) DNA segments in both samples are run in parallel for electrophoretic separation.

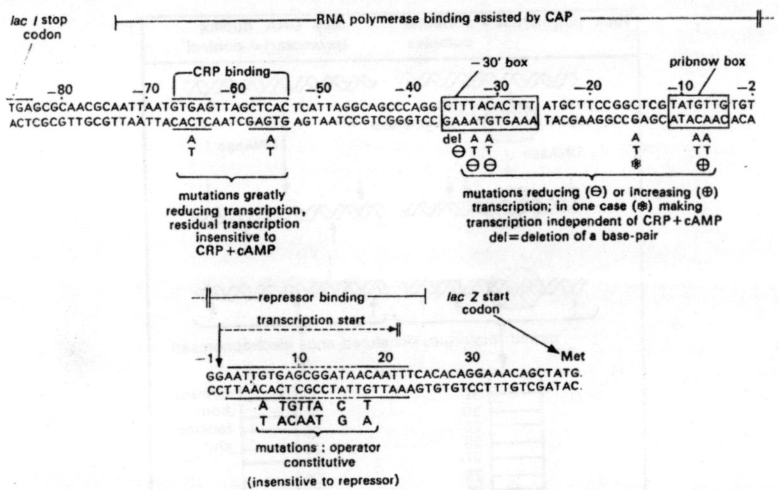

Nucleotide sequence of DNA in E. coli lac transcription control regions, showing functions of different regions and the effects of mutations in these regions.

A comparison of electrophoretic patterns in two samples will show a number of bands missing in the gel, where DNA-protein complex was used as a sample. The missing bands represent the position of DNA sites protected by protein and will allow the determination of DNA region used by protein for binding. If above samples are treated for sequencing, the sequence of binding site can be directly 'read off from the gel.

DNA Sequences for CAP, RNA Polymerase and Lac-repressor: Entire DNA segment has been sequenced from transcription termination (downstream) end of 'lac F gene to beyond the transcription start of Z, Y, A genes of lac operon.

This sequence contains the promoter and operator segments. Different parts of this sequence have been studied for their functions. For instance, sequences of three binding sites for CAP, RNA polymerase enzyme and 'lac' repressor could be identified through the technique of DNA footprinting. These sequences are shown in Fig.

Identification of Starting Point : In the DNA segment sequenced as above, starting point of transcription in *'lac'* operon could be identified with the help of S_1 nuclease. This enzyme has the ability to degrade single-stranded DNA or RNA, but not DNA-RNA hybrids. Therefore, a DNA restriction fragment (obtained due to treatment with a specific restriction endonuclease) containing the starting point and some transcription region of DNA, was hybridized with excess of lac mRNA (5' end of mRNA will be complementary to startpoint in DNA) and then digested with S_1 nuclease. The size of protected duplex DNA-RNA hybrid will determine the position of starting point (upstream) from the downstream restriction site. This sequence used as starting point is shown in Figure.

Pribnow Box and Other Sequences Common to DNA Regions Upstream to Several Operons : Upstream to starting point of lac operon are number of sequences which are common to many E. coli and bacteriophage genes. There is a sequence of about 10 bases upstream from starting point (also called minus ten or - 10 sequence) called Pribnow box which is an important sequence and named after D. Pribnow, who first discovered it in 1975. Pribnow box is shown in Figure and has the following sequence in lac operon.

-10 -5

TAT GTTG

ATA CAAC

This sequence differs slightly from the consensus sequence for Pribnow box (TATAAT), where it was shown that among prokaryotes the range of conservation of bases varies from 45% to 96%. There are other common sequences located at -30 and -60 bases (upstream), the latter being CAP site, which is marked by an interrupted three base pair inverted repeat, flanked by non-inverted T-A base pair on either side as follows.

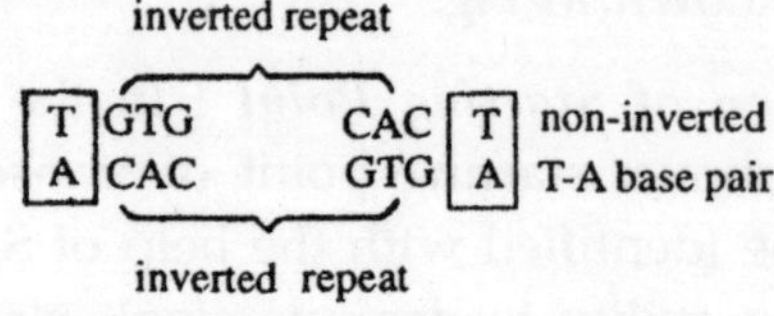

The TGTGC pentamer may lie in either orientation. This CAP binding site is common to *lac*, *gal* and *ara* operons and may lie at different locations relative to start point in various operons. In gal operon, this site may be within the promoter (between –53 to –23 bp), in lac operon this site lies between –72 to –52 bp and in ara operon it lies between –107 to –78 bp. Several of these above DNA sequences were found to be important, because mutations in these sequences influence promoter function (Fig.).

DNA Rearrangement Regulation

Expression of certain genes in prokaryotes are also regulated through DNA rearrangements (for DNA rearrangements in eukaryotes). These DNA rearrangements can be broadly classified into two groups.

Stochastic (random) Gene Rearrangements : These rearrangements take place without any need, so that a small fraction of the cell population carrying rearranged genes behaves like boy scouts, always ready (if the eventuality for which they are prepared does not occur, no problem). One of the best studied examples of these gene rearrangements involves inversion of the DNA elements carrying the promoter for one of the flagellar protein operons in Salmonella typhimurium. In one orientation, the element promotes transcription of a gene encoding one 'flagellar protein' and 'repressor' of a distant gene encoding an alternative flagellar protein. Inversion of the elements blocks transcription of both, the 'first flagellar protein gene' and the 'repressor gene' of the second alternative flagellar protein, thus permitting expression of the second flagellar protein.

Developmentally Regulated Gene Rearrangements : These gene Rearrangements take place during the differentiation of heterocysts (not in vegetative cells) in cyanobacteria or in the mother cell (not in forespore, which forms spores) of Bacillus subtilis during sporulation. Both the heterocyst (in cyanobacteria) and the mother cell (in B. subtilis) are terminal cells and do not give rise to cells of next generation, so that the gene Rearrangements are not transmitted. These developmentally-regulated gene Rearrangements involve excision of DNA segments that interrupt coding regions of the genes. Moreover, these gene Rearrangements are not random and require a series of signal transductions, gene activations and morphological differentiation.

Many species of filamentous nitrogen fixing cyanobacteria like Nostoc and Anabaena have specialized cells called heterocysts, where nitrogen fixation occurs during conditions of nitrogen limitation. Heterocysts occur at regular intervals making upto 10% of the cells in free living Anabaena and Nostoc, this frequency being much higher in symbiotic associations. The nitrogen fixation is facilitated by the presence of a 'nitrogenase complex' with three polypeptide I components, that are encoded by three linked genes, nifHDK. In Anabaena, nifK gene is 1 lkb downstream from NifHD genes. This 11kb element is excised and the *nifHDK* operon is reconstructed. A second excision removes a 55kb element, which interrupts the coding region of a gene called *fdxN*. Both excisions are catalysed by site-specific re-combinases encoded within the interrupting elements themselves. In the absence of these excisions, it is impossible to transcribe the genes downstream of the interrupting elements.

Another example of DNA rearrangement is found in Bacillus subtilis, where some of the mother cell-specific transcriptions (including those for spore coat proteins) require a form of RNA polymerase carrying σ^K (sigK, a 27kD protein). The sigK is encoded by sigk gene resulting from the DNA Rearrangements between two open reading frames (ORFs), spoIVB and spoIIIC (formerly believed to be separate genes). The DNA rearrangement, in this case, involves excision of a 42kb element. It was shown that the strains without this 42kb element are unable to sporulate (they sporulate if a copy of sigk gene is supplied) and the strains with this 42kb element sporulate normally. (For more details on DNA Rearrangements in prokaryotes consult Ann. Rev. Genet. 26 : 113-130, 1992).

Importance of Post-transcriptional Regulation

Leader Sequences and Attenuators : While there are mechanisms controlling transcription of DNA as discussed above, there are also mechanisms, which operate after the transcription is initiated or sometimes even when the transcription is over. For instance, in tryptophan operon of *E. coli*, the mRNA of gene trpE (gene for enzyme anthranilate synthetase) has a sequence of 162 bases between its initiation codon AUG and the promoter-operator region of this gene. This sequence of 162 bases is called 'leader' sequence and has several features of interest, which determine post-transcriptional control. (This post-transcriptional control is in addition to the normal transcriptional control provided in operon model). The leader sequence at its 3' end, but 8-12 bases before the initiating codon, has a sequence 5'AGGAGGA3', which is common to several E. coli genes and was discovered by J. Shine and L. Delgarno. This sequence 5'AGGAGGA3' (known as Shine Delgarno sequence) base pairs with the sequence 5'-ACCUCCUUA-OH-3' found at 3' end of 16S rRNA and thus helps in mRNA binding to ribosome.

Besides the sequence 5'AGGAGGA3' for ribosome binding,

there are other sequences of interest in the leader sequence. In leader sequence of trpE discussed above, a mechanism is available, which allows further transcription when synthesis is needed, and stops the transcription whenever the synthesis is not needed. For instance, when tryptophan is in short supply, so that its further supply is needed, the transcription will take place and when tryptophan is in adequate supply, so that its further supply is not needed, the transcription will stop. This is achieved by a remarkable mechanism, which becomes clear by a unique feature common to leader sequences of several operons each meant for synthesis of a specific amino acid, e.g., his, leu and val in *E. coli*. This unique feature includes the presence of two successive tryptophan codons at a specific site in trpE leader sequence. Similarly, in leader sequence of phenylalanine operon, there are three consecutive phenylalanine codons; in leader sequence of leucine operon, there are four consecutive leucine codons and in leader sequence of histidine operon, there are eight consecutive histidine codons. These regions of leader sequences each having consecutive codons for the same amino acid, for the synthesis of which the operon is meant, gives rise to mRNA for a small polypeptide about 15 amino acids long. Significance of this polypeptide synthesized by leader sequence has been explained as follows.

In tryptophan operon, it has been shown that when tryptophan supply is adequate, transcription stops at about

```
5'        10        20        30        40        50
pppAAGUUCACGUAAAAAGGGUAUCGACAAUGAAAGCAAUUUUCGUACUGAAAGG
    60        70        80        90        100
UUGGUGGCGCACUUCCUGAAACGGGCAGUGUAUUCACCAUGCGUAAAGCAAUC
  110       120       130       140       150       160
AGAUACCCAGCCCGCCUAAUGAGCGGGCUUUUUUUUGAACAAAAUUAGAGAAUAA
    170       180
CAAUGCAAACACAAAAACCGACU.....3'
```

The leader sequence of mRNA transcribed from E. coli trp (tryptophan) operon. Specific regions for signals are underlined. Probable ribosomes binding Shine-Delgarno sequences at 15-19 and 152-155 nucleotides; at 27-29 and 165-167. there is initiating codon AUG; two adjoining trp codons (UGG, UGG) at 54-59; sequence 114-134, underlined by broken line, is termination structure (attenuator) shown in Fig.

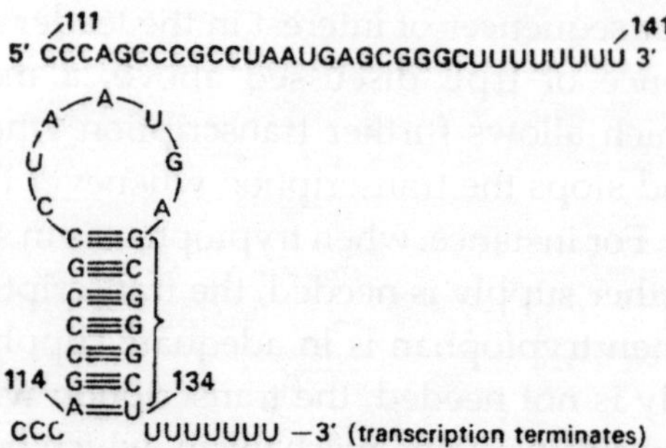

Termination structure (stem and loop), which is shown by broken underline in fig.

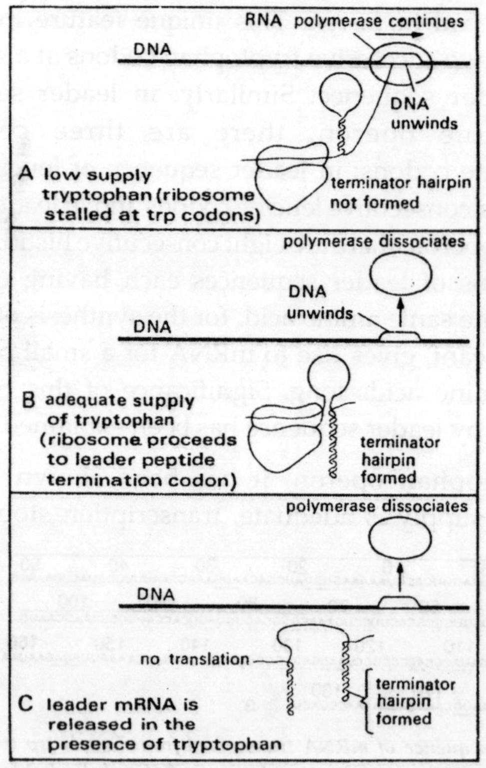

Different alternatives for RNA polymerase due to position of ribosome, which depends on hairpin formation.

147 bases after initiation, so that no for trpE and other genes of the operon are But if tryptophan is in short supply, transcription

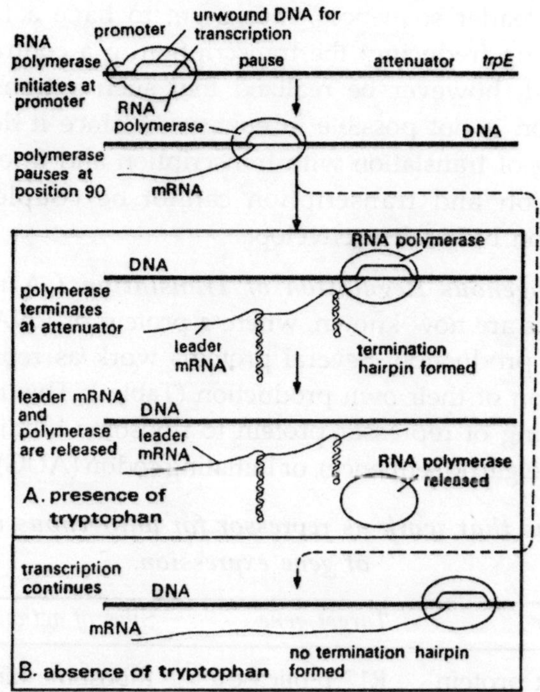

Role of attenuator in regulation of the transcription on tryptophan operon. (A) The presence of tryptophan allows formation of hairpin leading to release of RNA polymerase. (B) The absence of tryptophan allows the polymerase lo continue transcription (for details see text).

continues uninhibited. During adequate supply of tryptophan, mRNA for leader sequence is translated on ribosomes without any difficulty since two consecutive tryptophan codons mentioned above will be easily read. This favourable synthesis of leader peptide leads to the formation of hairpin loop, also known as attenuator, in the sequence between bases 114 and 134. This stops further synthesis of mRNA. On the other hand, if tryptophan is in short supply, the two trp codons (UGG, UGG) cannot be translated due to lack of availability of amino acid 'trp' and thus ribosome becomes stalled at this point (Fig.) and makes the formation of terminator hairpin loop (in sequence 114-134) impossible, thus indirectly permitting transcription of *trpE* and other genes (Fig.).

The leader sequences, thus seem to have a function of attenuating (reducing) the transcription in a controlled way. It should, however be realized that such a transcriptional regulation is not possible in eukaryotes since it depends on coupling of translation with transcription and in eukaryotes, translation and transcription cannot be coupled due to separation by nuclear envelope.

Autogenous Regulation of Translation : A number of examples are now known, where a protein or RNA regulates its own production. Several proteins work as repressors for repression of their own production (Table). This is achieved by binding of repressor protein to ribosome binding site (or Shine-Delgarno sequence), or initiating codon (AUG) of mRNA.

Proteins that work as repressor for autogenous regulation of gene expression.

Repressor	*Target gene*	*Sites of action*
R17 coat protein	R17 replicase	Ribosome binding site
T4 Reg A	T4 early genes	Initiation codon
T4 DNA *polymerase*	T4 DNA *polymerase*	Shine-Delgarno *sequence*
T4 p32	T4 gene 32	Single-stranded 5' leader

An important example of autogenous regulation is p32 protein, which is coded by T4 gene32 and needed for recombination, repair and replication of DNA. Whenever this protein p32 is in excess, it binds to its own mRNA and prevents initiation of translation. Several genes for ribosomal proteins, protein synthesis factors and RNA polymerase sub-units (which are organized in a few operons), are also autogenously regulated at the translation level. In these cases, a regulatory protein binds to a polycistronic mRNA and inhibits expression of a contiguous sets of genes within an operon, always including its own gene. This is true about ribosomal proteins (r proteins). These r proteins are generally utilized for binding with rRNA for ribosomes assembly, but if there is excess of r proteins, one of the r proteins may bind to polycistronic mRNA and inhibit synthesis of r proteins.

In the above cases, the mDNA remains intact but cannot be translated. However, there are other systems, where mRNA may be degraded as in case of synthesis of tubulin in eukaryotic cells. In this case tubulin molecules are normally assembled as microtubules, but when present in excess these may bind at a short specific sequence of its mRNA causing its degradation.

Control by Substitutive Splicing : We had discussed the phenomenon of alternative splicing, which means that a transcript or a precursor mRNA may follow one of the several paths of alternative splicing, so that the same transcript may lead to the synthesis of different protein 'isoforms' in time and space. This mechanism has been shown to be widespread in organisms ranging from viruses to mammals and is used for regulation of gene expression at the post-transcriptional level.

Regulation by Anti-sense RNA : During late 1980's and early 1990's, it has been shown that the translational control of protein synthesis can be exercised by using RNA, which is complementary to mRNA, so that this introduced RNA may form RNA-mRNA hybrids and prevents mRNA from being translated. Since this RNA will be complementary to mRNA and will interfere with its translation, it is called anti-sense RNA or mic RNA (mic = mRNA interfering complementary RNA). Such anti-sense RNA can be used to stop translation of mRNA in cells at will, and therefore is reported as a method analogous to production of mutations at will. But more important is the use of this phenomenon by cells to exercise control of gene expression at the level of translation.

In transposon Tnl0, translation leading to the production of transposase (a protein required for transposition) is inhibited through the production of anti-sense RNA. The production of anti-sense RNA is controlled by a promoter which directs transcription in opposite direction to the transposase gene, producing anti-sense RNA, which overlaps 36 bases of mRNA at its 5' end, and is complementary to it. There is definite evidence that the synthesis of transposase enzyme in Tn10 is regulated through the synthesis of anti-sense RNA.

In E. coli also, the role of anti-sense RNA in regulating gene expression has been demonstrated in case of two genes for outer membrane protein. These two genes are ompC and ompF (Omp = outer membrane protein). The total amount of these two proteins remains constant, although their relative quantities may differ according to requirement. It has been shown that a DNA fragment upstream of ompC when introduced into ompF+ cells, inhibited the production of OmpF protein. This DNA fragment coded for a small 174 base RNA and is termed mic RNA (mRNA interfering complementary RNA), which was transcribed in a direction opposite to ompC gene. Since no ribosome-binding site is present on mic RNA, it can not be used for protein synthesis. But it has extensive homology with 5' end of ompF mRNA, and therefore, can form a hybrid with it. Since the sequences present in mic RNA include sequences complementary to ribosome binding site of ompF mRNA, hybrid formation will inhibit the translation initiation. Such a mechanism is believed to result in coordinate regulation of the production of OmpC and OmpF proteins.

The above two genes ompC and ompF are known to be involved in controlling the response to changes in osmolarity. Another locus envZ codes for a receptor protein (EnvZ) that serves as an osmosensor. EnvZ activates another protein (OmpR), the product of ompR. The protein OmpR regulates the transcription of ompF and micF genes. Increase in osmolarity leads to synthesis of micF RNA, which turns off translation of ompF mRNA through complementary base pairing (anti-sense RNA mechanism).

Another example of the regulation of translation through anti-sense RNA is available in the nematode, Co-enorhabditis elegans. In this case, expression of the gene Iin4 (regulating larval development) is controlled by another gene Hnl4 coding for two transcripts (22 and 61 nucleotides long) having complementarity with a sequence transcribed from *lin4*. It is possible that these transcripts derived from *lin14* work as anti-sense RNA molecules.

Even in eukaryotes, it is suggested that anti-sense RNA (or even anti-sense DNA) may be used to inhibit gene expression. The presence of complementary RNA (or DNA) sequence was actually shown to reduce gene expression. The extent of utilization of this mechanism for control of gene expression in cell is not known, but it is being definitely utilized now for artificial manipulation of gene expression.

Inhibition of cloned genes by anti-sense RNA (or DNA) technology has been achieved in several cases, using recombinant-DNA technology through the construction of anti-sense expression vectors. These genes whose expression was thus manipulated, included the following : (i) thymidine kinase (TK) gene from herpes simplex virus (HSV), (ii) actin gene (actin protein is a major component of cytoskelston), (iii) genes for pigment synthesizing enzyme in petunias; (inhibition of this enzyme led to unusual pigmentation in flowers), (iv) genes for enzymes responsible for fruit ripening in tomatoes (inhibition of these genes will help in slow ripening) and (v) plant virus genes causing diseases. In all these cases, an anti-sense expression vector was constructed, so that the orientation of the gene of interest is reversed with reference to promoter sequence. These will lead to the synthesis of anti-sense RNA inhibiting the translation of mRNA. Using this technology inhibition of the expression of several genes as above was achieved successfully.

Not only anti-sense RNA, but artificially synthesized

oligonucleotides (both DNA and RNA) have also been used to inhibit the translation of I mRNA. This technology will be increasingly used in future for study of gene function and for treatment of diseases.

***Repression and Activation of Translation* :** The number of regulatory circuits identified at the translational level increased steadily during 1980's and 1990's. Besides, the mechanisms of regulation discussed above, control at the translational level is also achieved by formation of secondary and tertiary structures, which either inhibit or facilitate translation. These secondary or tertiary structures are dependent on intramolecular base pairing in mRNA, and can be variously modified due to effect of molecules acting on specific target sites. The regulatory circuits at the translational level may either involve repression or activation.

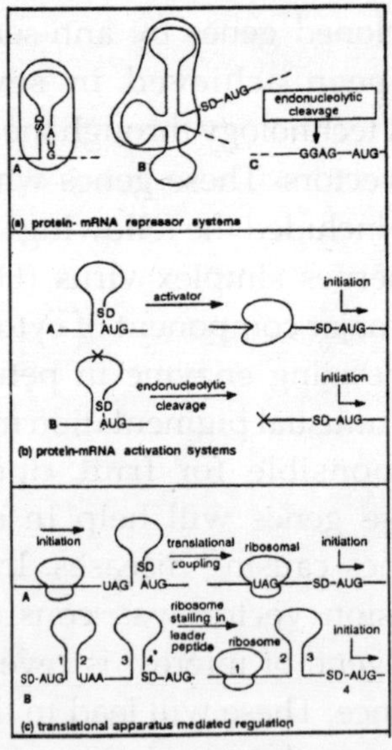

Some models of translational regulation in prokaryotes (repression and activation; see text for details).

***Repression of Translation* :** At least three different situations of repression of translation have been proposed (Fig.) : (i) A repressor-effector molecule (e.g. phage R17 coat protein) may recognize and bind to a specific sequence or to a specific secondary structure (involving SD region and AUG codon), thus blocking initiation of translation, through blocking of the ribosomal binding region, (ii) A repressor effector molecule may bind to an operator (not involving SD region and AUG codon), thus stabilizing an inhibitory mRNA secondary structure (e.g. threonyl-tRNA synthetase; phage T4 RegA protein, which recognizes TIR sequence domains of a number of T4 genes), (iii) An effector molecule (an endonuclease) can inhibit initiation of translation by endonucleolytic cleavage of SD region (e.g. phage T4 motA mRNA SD region).

Activation of Translation: Cases of translational activation have also been reported, even though they are less well understood. The most likely mode of action of these positive effectors or activators is through destabilization of inhibitory secondary structures in mRNA either through simple binding or by endonucleolytic cleavage (Fig., e.g. phage X and phage T4 RNase III). Translation of certain genes is also influenced by the translation of neighbouring genes, a phenomenon described as translational coupling. Under this mechanism, the ribosome terminating the translation on a preceding gene of a polycistronic mRNA, may sometimes release the inhibitory secondary structure found in the mRNA of the following gene (e.g. 6-phosphogluconate dehydrogenase gene and erythromycin methylase gene; see Fig.).

Feedback Inhibition

In some cases, the end product of a particular biosynthetic pathway gets accumulated if not utilized. Sometimes such an accumulation may stop further synthesis of this substance. For instance, if E. coli cells are supplied with amino acid isoleucine, it blocks activity of enzyme involved in the first step of isoleucine biosynthesis. This inhibition is due to binding of isoleucine to enzyme threonine deaminase (Fig.). This mechanism of regulation of enzyme activity is known as feedback inhibition. If the first enzyme is blocked in this case, it may not be necessary to block the activity of other intermediate enzymes, since there will be no accumulation of intermediate products.

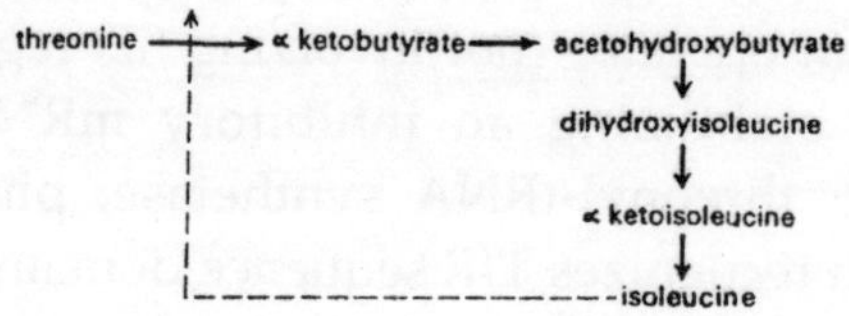

Feedback inhibition in the synthesis of isoleucine in E. coli.

The end product in the above case perhaps acts through allosteric transformation of the enzyme, as a result of which the site of activity of the enzyme gets transformed in such a way that it cannot perform its catalytic function. This would mean that in feedback inhibition the enzyme does not function, while in repression no mRNA synthesis and hence no enzyme synthesis takes place. Therefore, this is not a case of control of gene activity, but gene expression is controlled only at a post-translational level.

Signal Transduction and 'Two Component Regulatory Systems' : In prokaryotes, a wide variety of stimuli are sensed and transduced leading to altered gene expression. Important examples in E. coli, include chemotaxis, nitrogen assimilation and outer membrane protein expression. Two sets of proteins, bearing sequence similarities, are known to be involved in these regulatory systems, which are, therefore, described as 'two component regulatory systems'. The first component is a protein kinase, described as 'kinase', that auto-phosphorylates (using ATP) a histidine residue,' and transfers its phosphoryl group to an aspartate residue of the second component described as 'regulator'. Finally, a phosphatase activity may reset the system by restoring the regulator to its unphosphorylated state. Regulator phosphate levels are controlled in response to environmental condition. The above signal transduction, using 'two component systems' can be illustrated using the following three examples.

Bacterial Chemotaxis : In homogenous environment, E. coli swims smoothly, but it moves towards or away from wide variety of compounds classified as 'attractants' or 'repellents' respectively. This mechanism involves a post-translational regulation of gene expression, involving reversible phosphorylation of two .proteins, CheA (kinase) and CheY (regulator).

Nitrogen Assimilation : Many species of bacteria regulate both the amount and the activity of 'glutamine synthetase' enzyme in response to nitrogen availability. The gene for this enzyme is glnA and its major promoter is glnAp2. Transcription of the gene is regulated by a two component system (NtrB = kinase, NtrC = regulator). NtrB auto-phosphorylates and transfers its phosphate group to NtrC, which in its phosphorylated form activates glnAp2. The phosphorylation of NtrC is regulated by nitrogen availability through another protein called 'PII'.

Outer Membrane Protein Expression : Earlier in this chapter, we discussed regulation of expression of genes ompC and ompF (meant for outer membrane proteins) at the translational level using anti-sense RNA. Another mechanism at the transcriptional level involves a two component system (EnvZ-kinase; OmpR = regulator). EnvZ auto-phosphorylates and transfers its phosphate group to OmpR, which in its phosphorylated form, stimulates expression of ompF under 'low osmolarity' conditions and that of ompC under 'high osmolarity' condition. Phosphorylation of OmpR is also regulated due to hydrolysis in the joint presence of EnvZ and ATP.

signal transduction, using two component systems can be illustrated using the following three examples.

Bacterial Chemotaxis : In homogeneous environment, E. coli swims smoothly, but it moves towards or away from wide variety of compounds classified as 'attractants' or 'repellents' respectively. This mechanism involves a post translational regulation of gene expression, involving reversible phosphorylation of two proteins, CheA (kinase) and CheY (regulator).

Nitrogen Assimilation : Many species of bacteria regulate both the amount and the activity of glutamine synthetase enzyme in response to nitrogen availability. The gene for this enzyme is glnA and its major promoter is glnAp2. Transcription of the gene is regulated by a two component system (NtrB = kinase, NtrC = regulator). NtrB auto phosphorylates and transfers its phosphate group to NtrC, which in its phosphorylated form activates glnAp2. The phosphorylation of NtrC is regulated by nitrogen availability through another protein called 'PII'.

Outer Membrane Protein Expression: Earlier in this chapter we discussed regulation of expression of genes ompC and ompF (meant for outer membrane proteins) at the translational level using antisense RNA. Another mechanism at the transcriptional level involves a two component system (EnvZ = kinase; OmpR = regulator). EnvZ auto-phosphorylates and transfers its phosphate group to OmpR, which in the phosphorylated form stimulates expression of ompF under low osmolarity conditions and that of ompC under high osmolarity conditions. Phosphorylation of OmpR is also regulated due to hydrolysis in the joint presence of EnvZ and ATP.

3

Synthesis of Genes

Molecular manipulation of specific genes is one of the popular areas of research in biotechnology. These genes, therefore, should be either isolated or artificially synthesized before they are manipulated and used for transformation leading to the production of transgenic animals and plants. The isolated genes are also sometimes sequenced for a better understanding of its structure, that needs to be manipulated. We, therefore, discuss in this chapter the gene technology available for isolation, sequencing and synthesis of specific genes.

Genes' Isolation

During the last three decades (1970-2000), significant progress has been made in the techniques for isolation of a variety of genes, including those for (i) ribosomal RNA; (ii) specific protein products; (iii) phenotypic traits with unknown product and those for (iv) regulatory functions e.g. promoter genes, etc. Different techniques have been used for the isolation of these different types of genes and will be briefly described later.

Early Effort in Isolating Ribosomal RNA Genes in Xenopus: As we know, ribosomes consist of ribosomal RNA (rRNA) and proteins. Ribosomal RNA (rRNA) can be of four sizes namely, 28S, 16S, 5.8S and 5S (S stands for Svedberg units, a measure of speed with which a molecule sediments on centrifugation). This ribosomal RNA makes 80 per cent of cellular RNA and is synthesized from ribosomal genes, which could be isolated. Isolation of ribosomal genes was considered convenient due to their following three characteristic features : (i) availability of homogeneous rRNA; (ii) differences between ribosomal RNA genes and other genes, due to relatively high G+C content in rRNA (rRNA has 45-60% G+C, while the remaining RNA has only 40% G+C); (iii) ribosomal genes are present in multiple copies, their number within a cell sometimes approaching several thousands.

In view of the above, ribosomal RNA genes were isolated for the first time (in 1965) in an amphibian named *Xenopus*, by Hugh Wallace and Max L. Birnstiel, then working at the University of Edinburgh (U.K.). Following steps were involved in this isolation: (i) rRNA was isolated from ribosomes of *Xenopus* and made radioactively labelled, due to its replication in a medium containing tritiated uridine; (ii) ribosomal DNA was isolated by density- gradient centrifugation followed by its denaturation (G+C content of rDNA differs from that of bulk DNA and helps in its separation by centrifugation); (iii) single-stranded DNA was fixed on a filter paper; (iv) labelled rRNA was added on filter paper carrying single-stranded DNA; (v) DNA-RNA hybridization was allowed to take place; (vi) excess-labelled RNA was washed; (vii) radioactivity was measured and duplex hybrids isolated, which on denaturation, gave single-stranded DNA, which could be made double stranded.

Isolation of ribosomal DNA allowed the first characterization of ribosomal genes, which was followed by the isolation of 5S genes by Donald Brown. For a detailed account of isolation of ribosomal genes in *Xenopus* as originally done, readers are referred to an article Isolation of Genes in Scientific American (August, 1973).

Isolation of Genes Coding for Proteins of Specific Nature: Isolation of genes for specific proteins became possible only after the discovery of reverse transcriptase enzyme in 1970. This enzyme can be easily used for the synthesis of copy DNA or complementary DNA (cDNA) from mRNA. This complementary DNA can then be used for the isolation of the corresponding gene from genomic DNA. It is, therefore, obvious that for the isolation of a specific gene, techniques should first be available for the isolation of specific mRNA. For this purpose, antibodies are produced against a specific protein for which the gene is to be isolated.

Therefore, isolation of a gene coding for a specific protein involves the following basic steps (Fig.): (i) purification of the protein product of the gene; (ii) production of antibodies against this protein product by immunizing animals like rabbit or mouse: (iii) precipitation of polysomes engaged in synthesizing specific protein with the help of the antibody raised in step (ii); (iv) isolation and purification of mRNA from the polysome fraction; (v) using mRNA for synthesizing cDNA with the help of reverse transcriptase; (vi) cloning of cDNA thus synthesized for preparation of a cDNA library; (vii) immunological and electrophoretic analysis of the translation products of cDNA clones to identify specific cDNA clone meant for the specific protein in mind; (viii) use of the specific cDNA probes selected in step; (vii) for the identification and isolation of the gene from genomic DNA through screening a complete or partial genomic library (for preparation of cDNA and genomic DNA libraries.

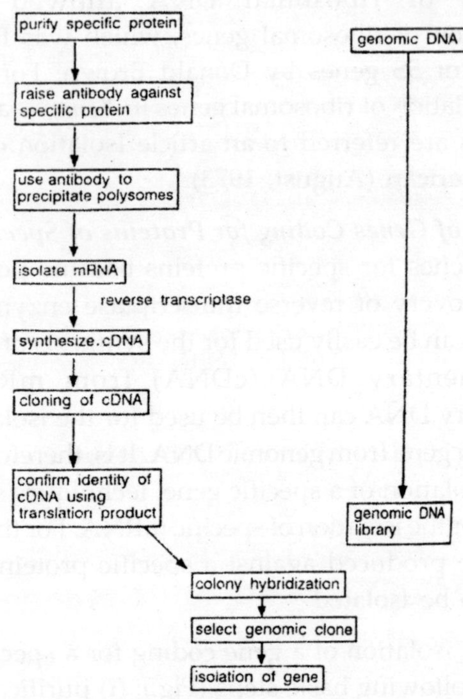

Isolation of a gene for a specific protein.

In step (vi) above for preparation of cDNA library, cDNA may be cloned in expression vectors (vectors in which gene can be expressed due to the presence of promoter sequences for RNA polymerase enzyme) like λ gt ll, which can accept gene insertions into β galactosidase gene. The chimeric vector will then produce hybrid protein if correctly expressed. The hybrid protein can be identified using antibody as above. Plaques belonging to the chimeric vector carrying the desired gene can also be identified by the reaction of antibody attached to a radioactive protein.

The above technique for isolation of gene has now been successfully utilized both in plants and animals. However, the first genes isolated were those which existed in multiple copies and were expressed at high levels in specific tissues, e.g. ovalbumin gene in chicks, globin and immunoglobulin genes in mouse, genes for storage proteins in cereals and legumes, amylase genes in barley, actin genes in some legumes, etc.

Genes' Isolation

(with known or unknown products) having tissue-specific expression

It is much easier to isolate genes which are expressed in specific tissues. For instance, genes for storage proteins are expressed only in developing seeds, ovalbumin gene is expressed in oviduct or globin gene is expressed in erythrocytes. Such genes can be easily isolated because mRNA extracted from these specific tissues will either exclusively belong to the gene of interest or it will be rich in this species of mRNA. Other mRNA molecules in minor quantities can be eliminated, since these can be identified through hybridization with mRNA from tissues where this gene is silent.

This strategy which is outlined in Figure, was actually followed for isolation and cloning of carrot genes expressed during the development of somatic embryos, and also for the isolation of several genes for storage proteins in crop plants.

Genes' Isolation

(with known or unknown products) using DNA or RNA probes

Specific molecular probes (whether DNA or RNA probes), if available can be used for isolation of specific genes. These probes may be available either from another species for the same gene or may be artificially synthesized using a part of the amino acid sequence of the protein product of the gene in question.

Use of Heterologous Probes : The probes obtained from one species and used for another species are called *heterologous probes.* These heterologous probes have been found to be effective in identifying gene clones during colony hybridization or plaque hybridization or on Southern blots. For instance, the gene for *chalcone synthase* (CHS) has been isolated from *Antirrhinum majus* and *Petunia hybrida* using heterologous cDNA probe from parsley. Similarly, heterologous *Antirrhinum* cDNA probe was used for isolation of CHS gene from barley and heterologous probes from maize were used for isolation of barley genes, *Wx* (waxy gene) and *Al* (aleurone gene). Heterologous probes should ordinarily be used with cDNA library and not with the genomic library, since in the latter case, unrelated genes or pseudogenes (which do not express) may be isolated and cloned. These heterologous clones if available in expression vector (e.g. pGEM-4) can also be used for getting RNA probes, which have been found to be more sensitive and efficient.

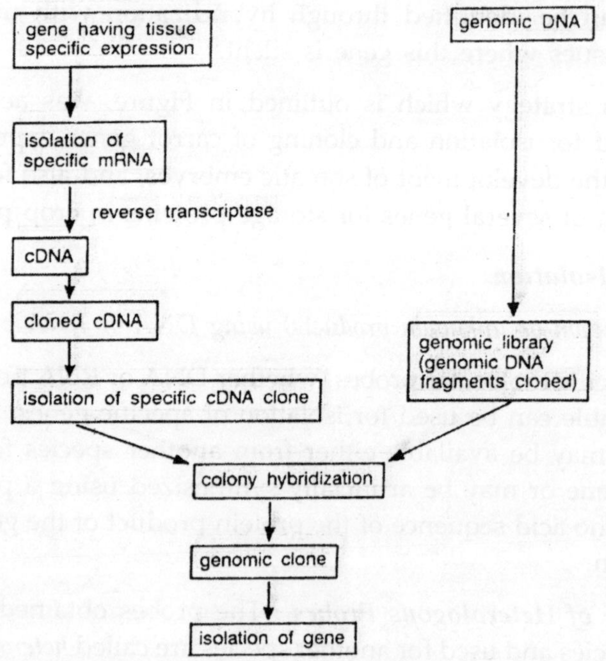

Isolation of a gene with tissue specific expression using its mRNA.

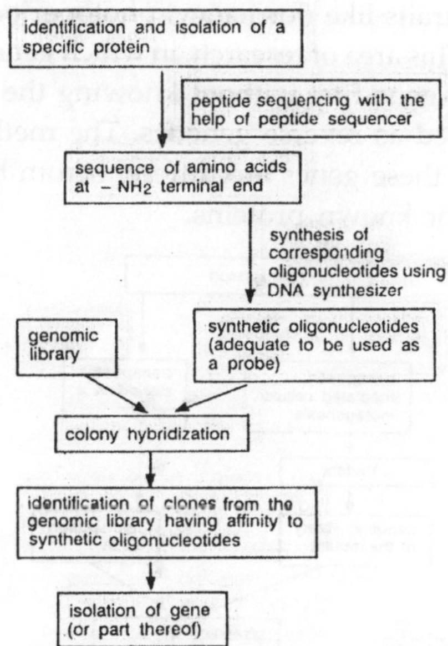

Isolation of a gene using the probe, artificially synthesized, on the basis of amino acid sequence of a part of the protein.

Use of Synthetic Probes : If protein, purified using the technique of two-dimensional gel electrophoresis, is used for micro-sequencing of 5-15 consecutive amino acids, this information can be used for the synthesis of oligonucleotides (using automated DNA synthesizers). These oligonucleotides may then be directly utilized for screening of cDNA or genomic libraries for isolation of specific genes. These can also be used for designing primers for the amplification of a gene in DNA extracted from a tissue. A 32 kilodalton (kD) glycoprotein (isolated from style) associated with the S_2 incompatibility allele of *Nicotiana alata* was used for synthesis of an oligonucleotide leading to the isolation of a related gene.

Isolation of Genes Coding for Unknown Products, When Probes are not Available : In some cases, we are interested in isolation of a gene whose phenotypic effect is known, but the gene product has not been identified or cannot be isolated, and no probe is available. Such genes include those for morphological traits like dormancy, photoperiodicity, disease resistance, etc. This area of research, in which genetics is studied by isolating the gene first without knowing the gene product is often described as reverse genetics. The methods used for the isolation of these genes are different from those used for genes coding for known proteins.

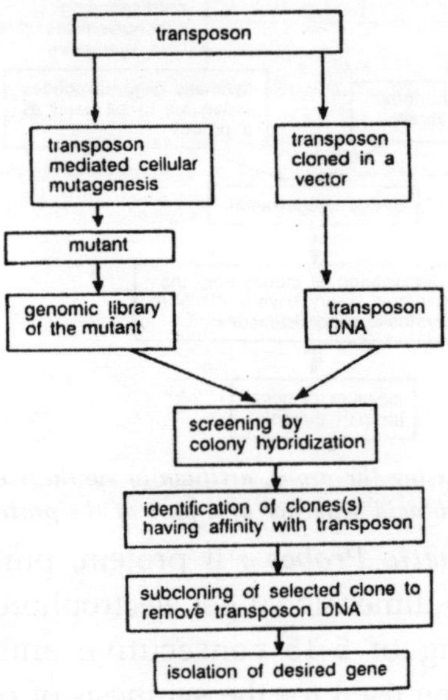

Use of transposon mutagenesis (or tagging) form identification and isolation of a gene.

Employing Transposable Factors (Transposon Tagging) : The transposable elements (TEs), in some cases, have been effectively utilized for isolation of genes, when the gene product is unknown. In this case a transposon works as a gene tag. Following steps which are illustrated in Figure, are involved in this procedure (i) clone a known gene with a scorable phenotypic effect; (ii) TE is transposed to this gene to get an unstable allele; (iii) this unstable allele is cloned and TE is isolated from this unstable allele (this is to select a TE which can produce unstable allele); (iv) this TE is transposed to a gene of interest with known phenotypic effect, to produce unstable allele; (v) the DNA is extracted from this mutant; (vi) TE sequence is used as a probe to isolate and clone the mutant gene (carrying inserted TE), so that we can then isolate the gene of interest.

Genes Isolated using Transposon Tagging in Plants

Gene	*Gene function*	*Species*	*Transposon (T_n)*
Al	Enzyme required for anthocyanin biosynthesis	Maize	*Mac En/Spm*
$B_{,}l$	Enzyme required for anthocyanin biosynthesis	Maize	*Ac*
B_z2	Regulatory gene	Maize	*Ds; Mu*
Cl	Enzyme required for anthocyanin biosynthesis	Maize	*En/Spm*
C2	Regulatory gene	Maize	*En/Spm*
P	Regulatory gene	Maize	*Ac*
R	Regulatory gene	Maize	*Ac*
Vpl	Leaf development	Maize	*Mu*
knotted	Regulatory gene	Maize	*Ds*
Opaque-2	Chloroplast development	Maize	*Ac- En/Spm*
hef-106	Enzyme required for anthocyanin biosynthesis	Snapdragon	*Mu*

In maize, TE like *Ac/Ds En/Spm*, and *Mul* have been isolated using genes *Wx, C2* and *Adhl.* Similarly, TEs like *Tam3* and *Tam7* have been isolated from snapdragon *(Antirrhinum majus).* These TEs have been used for gene-tagging experiments leading to isolation of genes. In maize, several genes like *Bzl, P, Al, Cl* and *C2* have also been isolated successfully using transposon tagging (Table).

For transposon tagging, often transposable elements endogenous to species, like maize and snapdragon have been used. However, rarely transposons available from one plant species can be moved into the genome of another plant species, whose gene is to be isolated. For instance *Ac* element of maize has been transferred to tobacco, where it can integrate into any locus permitting transposon tagging and gene isolation.

T-DNA Insertion Mutagenesis For Isolation of Plant Genes: A technique similar to transposon tagging is T-DNA insertion for isolation of genes (Fig.). A limitation of transposon tagging is that it is often confined to either the self fertilized diploid plants such as snapdragon or the cross-fertilized diploid crops like maize, both having well-characterized endogenous transposable elements (TE). Although derivatives of maize *Ac* have been successfully used as TE in crops to which they do not belong, their use is still limited and cannot be extended to all plant species. Therefore, T-DNA tagging involving random insertion of T-DNA into the genome, during *Agrobacterium tumefaciens* mediated transformation has been exploited. The most extensive use of T-DNA insertion mutagenesis has been in *Arabidopsis*, where 35-40% mutations in cultures were found to be tagged by T-DNA inserts. In 1995, already there were as many as 14,000

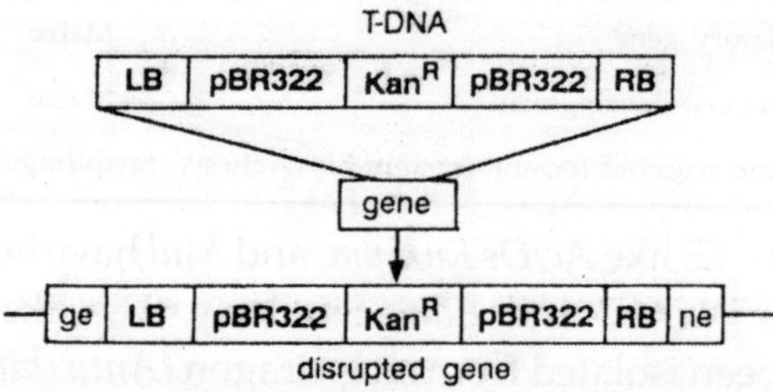

Use of T-DNA insertion for identification and isolation of a gene.

(d) gene trap- reporter gene preceded by a splice acceptor (SA) when inserted in an intron, a fusion protein is produced giving a mutant phenotype transformants, with an average of 1.5 inserts per transformant were available at the University of Arizona, Tucson (USA), 40% of them being due to T-DNA insertion. If *Arabidopsis* genome is considered to be 130 Mb and T-DNA insertion is random, there is 50% probability for any individual gene to be tagged among the above 14,000 transformants. A number of these genes have been cloned using T-DNA insertion and are being utilized for improvement of a variety of traits in a number of crop plants.

Promoter, Enhancer and Gene Trap for Isolation of Genes. Although transformants for a large number of traits have already become available, all genes are not tagged in these transformants. Therefore, a major effort is needed to isolate any new gene that is not tagged in any of the available transformants. A large number of transformed plants will have to be produced and propagated to isolate these new genes.

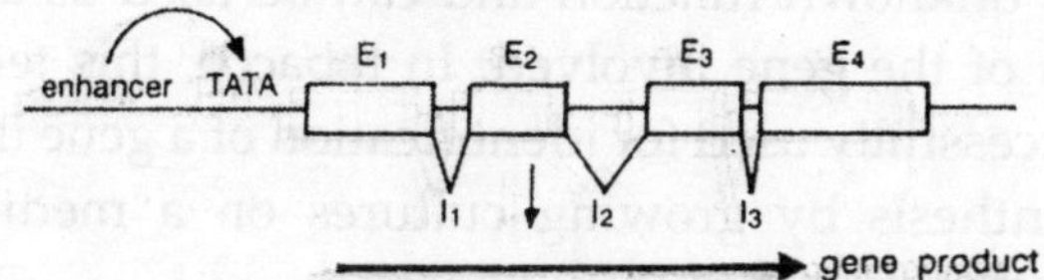

(a) a hypothetical gene – intron sequences are excised after transcription ($E_1 - E_4$ = exons ; $I_1 - I_3$ = introns)

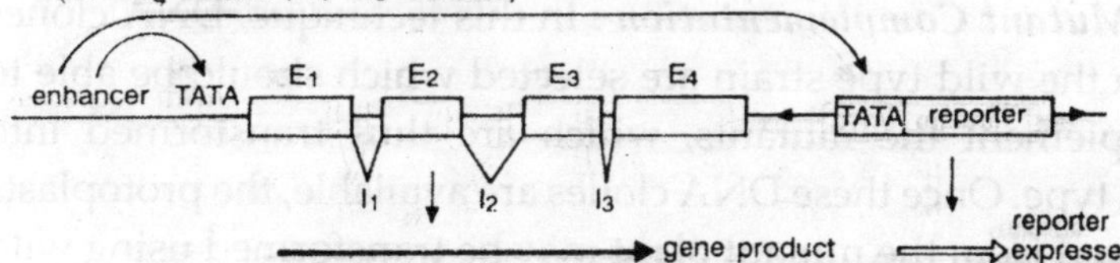

(b) enhancer trap – reporter gene with minimal promoter (TATA) is expressed, when inserted near a gene with enhancer

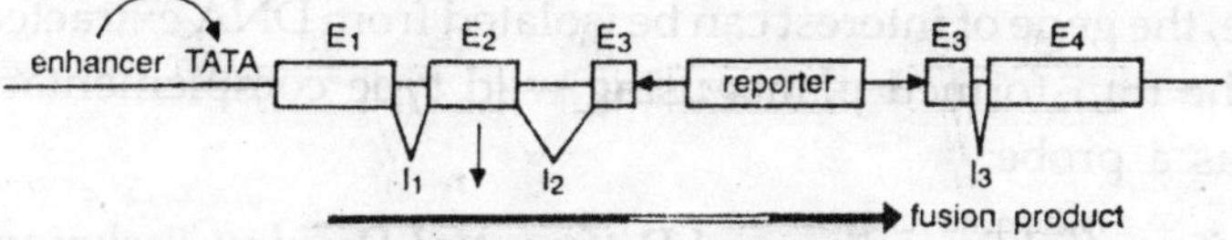

(c) Promoter trap – reporter gene without TATA when inserted in an exon, a fusion protein will be produced

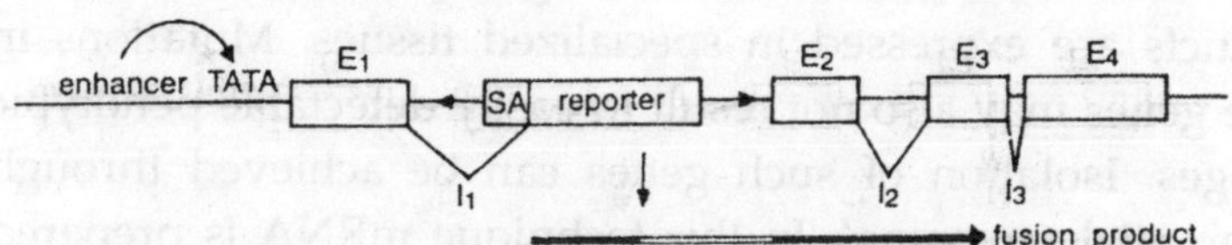

(d) gene trap– reporter gene preceded by a splice acceptor (SA) when inserted in an intron, a fusion protein in produced giving a mutant phenotype

Promoter, enhancer and gene trap strategies for identification and isolation of genes with unknown functions.

Furthermore, the insertion of transposon or T-DNA may not result in perceptible phenotypic change. Therefore, if a new gene (for which mutation could not be produced so far through transposon tagging or T-DNA insertion) needs to be isolated, one may like to use a new strategy to clone the gene of interest. One such strategy involves the insertion of a reporter gene with or without a TATA box or a splice acceptor (SA). In Figure, using a hypothetical gene (Fig.), three such strategies are illustrated, which are described as 'enhancer trap', 'promoter trap' and 'gene trap'. In each case, the reporter gene will be expressed under the influence of enhance™ promoter sequence of a gene of unknown function and can be used as a tag for the isolation of the gene involved. In tobacco, this technique has been successfully used for identification of a gene involved in auxin synthesis by growing cultures on a medium not containing auxin.

Mutant Complementation : In this technique, DNA clones from the wild type strain are selected which should be able to complement the mutants, which are thus transformed into wild type. Once these DNA clones are available, the protoplasts derived from the mutant plant may be transformed using wild type clones and the transgenic plants are produced. Once this is done, the gene of interest can be isolated from DNA extracted from the transformed plants using wild type complementary clone as a probe.

Differential Screening and Differential Display Technique for Isolation of Genes : A number of genes with unknown gene products are expressed in specialized tissues. Mutations in these genes may also not result in easily detectable penotypic changes. Isolation of such genes can be achieved through 'differential screening'. In this technique mRNA is prepared from contrasting tissues of plants such as the following :

(i) tissues/plants exposed to different environmental conditions (e.g., drought);

(ii) tissues, which differ in function,

(iii) plants at different developmental stages.

In all these cases, two samples of mRNAs derived from

two tissues or two plants should differ and the mRNA species which is not common in both the samples will be identified through differential screening using the following steps (Fig.)

(i) mRNA is used for synthesis of cDNA using reverse transcriptase;

(ii) cDNA thus obtained from both samples of mRNA are labelled and used for sequential probing of a genomic DNA library;

(iii) clones that are more strongly hybridized with the cDNA from one mRNA sample than with that from another contain genes that are differentially expressed in the samples.

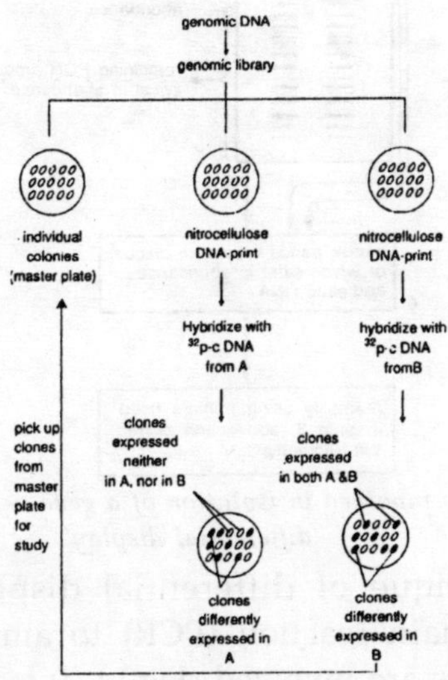

Different steps involved in isolation of a gene using the technique of 'differential screening'.

One of the limitations of differential screening is that the technique can be used for only those genes that have high expression, and may not prove suitable for those differentially expressed genes that have low expression (low abundance of mRNA). For such genes, another technique called 'differential display' is used.

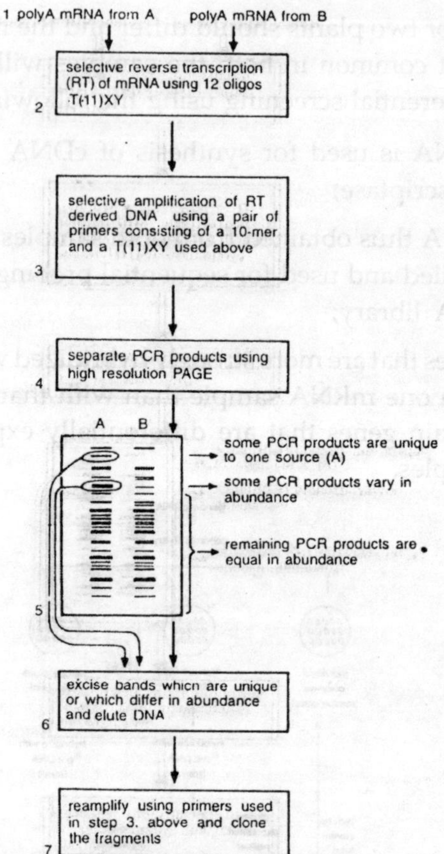

Different steps involved in isolation of a gene using the technique of 'differential display'.

The technique of differential display makes use of polymerase chain reaction (PCR) to amplify rare cDNAs. Following steps are involved (Fig.) : (i) isolate polyA mRNA from two sources as above (call them A and B) and run the following reactions separately with each of the two samples; (ii) conduct RT-PCR (reverse transcribed-PCR) ' using as primers 12 possible oligonucleotides of sequence. T(11)XY (XY stand for any two of the four bases, thus giving 12 possible dinucleotides excluding four homodinucleotides); the dinucleotide XY will select the mRNAs, thus providing a means of differential amplification; (iii) the RT-PCR product is amplified using a pair of primers, one of them being a random 10-mer and the other being the oligonucleotide used earlier in RT-PCR. This will allow amplification of only a fraction of RT-PCR products, including only those to which random 10-mer anneals, (iv) the PCR products obtained are separated on polyacrylamide gel electrophoresis (PAGE); it will be noticed that most of the PCR products are present in equal amounts in both samples but some will differ in abundance and still others will be unique in one sample (being absent in the other sample); (v) the unique bands or the bands showing differential expression are excised and reamplified using the same pair of primers as used in step (iii) above. The reamplified product is cloned and used as a probe for screening cDNA or genomic DNA libraries.

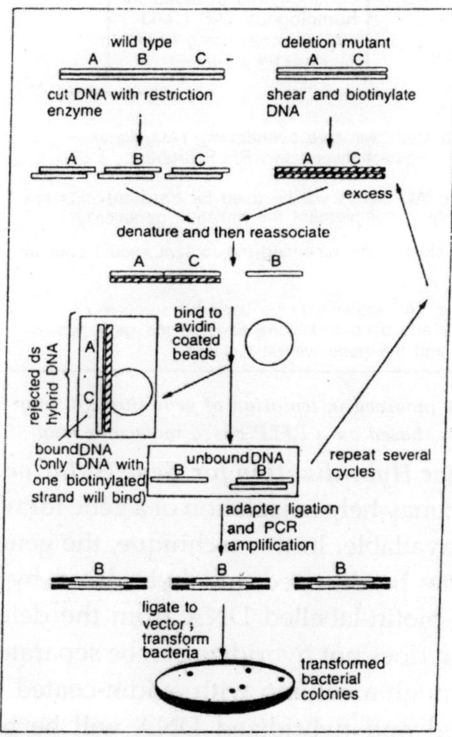

Different steps involved in isolation of a gene using the technique of 'subtractive hybridization'.

Since several genes may have similar patterns of expression in two tissues or two plants sampled, additional criteria will be needed to confirm the identity of isolated gene; this can be achieved by construction of transgenic plants expressing an anti-sense gene construct.

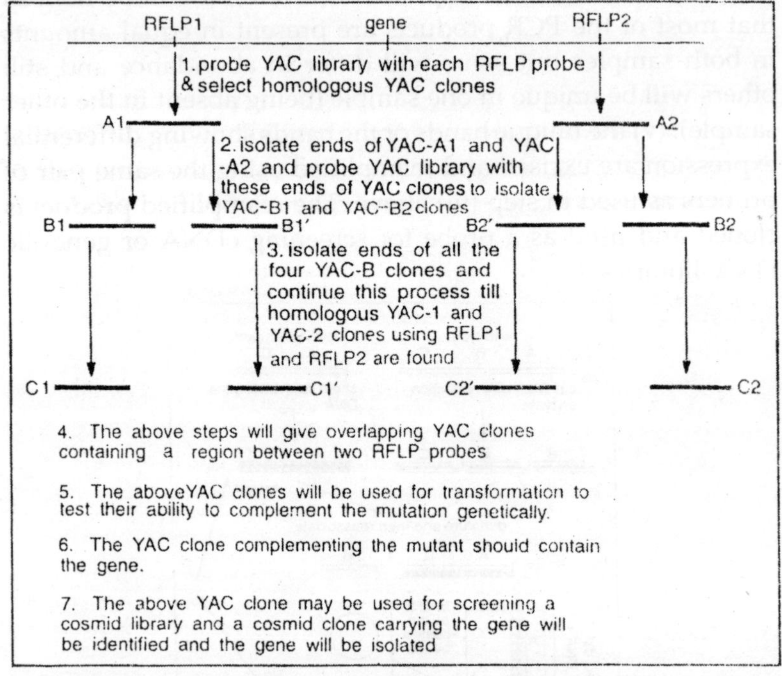

Different steps involved in isolation of gene through 'map based cloning' based on a RFLP based molecular map.

Subtractive Hybridization for Gene Isolation : Subtractive hybridization may help in isolation of a gene for which deletion mutants are available. In this technique, the genomic DNA of wild type may be digested and hybridized by an excess of sheared and biotin-labelled DNA from the deletion mutant. The DNA that does not hybridize can be separated by passing the DNA through a column with avidin-coated beads so that biotin labelled and hybridized DNA will be bound on the beads and the unlabelled unbound DNA will pass out from the column. This process may be repeated through several cycles. The eluted DNA is amplified using PCR (adapters of known sequence may be ligated to eluted DNA and these sequences may be used for designing primers) and then cloned (Fig.). The clones are then used on the southern blots of the mutant and wild type to select clones that hybridize only to the wild type" and not to mutant type. Further confirmation may be needed to identify the isolated gene.

Map-Based Cloning for Gene Isolation : Map-based cloning can be used for isolation of a gene for which a mutation can be identified and in a crop for which a saturated molecular map is available. Availability of comprehensive YAC or BAC genomic library is another requirement. Most of these requirements are met in major crop plants. The mutation is first mapped close to any of the molecular markers that are already mapped. A variety of molecular markers are available for the preparation of these molecular maps (e.g., RFLP, RAPD, SSR, STMS, AFLP, SNP, etc.). Once the mutant is mapped, the most closely linked pair of flanking markers are used as hybridization probes to isolate clones containing the genomic segment located between the markers. Techniques of chromosome walking, chromosome jumping and chromosome landing can be used to reach closer to the gene of interest and to isolate clones that may contain the gene to be isolated (Fig. for details of chromosome walking and chromosome jumping). These probable clones can be used for transformation of the mutant to find out which of the clones complements the mutation. This clone then should contain the gene. Using this approach, several genes have been isolated in crop plants and animals including humans. A gene for cystic fibrosis in humans was cloned through map-based cloning utilizing the technique of chromosome jumping. This technique will be increasingly used in future for isolation of a variety of genes for which the gene products are not known.

Novel Genes Isolation

In the year 1992, Prof. Asis Datta of JNU, New Delhi was selected for the Birla Award for Science and Technology for cloning and characterization of the following two novel genes, relevant to human health.

Gene for Oxalate Decarboxylase : In 1991, Prof. Datta's laboratory reported for the first time the cloning and characterization of the oxalate decarboxylase gene, which degrades and will thus reduce in our crop produce the content of oxalate or oxalic acid, which is harmful in many ways including the following : (i) Several green leafy vegetables (e.g. *Amaranthus*, spinach, etc.) being rich in oxalate, if ingested in large quantity are toxic, since oxalate chelates calcium and also destroys renal tissue in kidney, (ii) The attack and spread of a fungus *(Whetzelinia sclerotium)*, which causes damage to crops like sunflower involves accumulation of oxalic acid in the infested tissue, (iii) Oxalic acid is an essential substrate for the synthesis of the neurotoxin called p-N-oxalyl I-L-a, P diamino-propionic acid (ODAP), found in different parts of the plants of *Lathyrus sativus* (khesari dal). Consumption of this legume causes neurolathyrism, a well known disease causing damage to leg muscles, etc. ODAP interferes with metabolism of glutamic acid, which is involved in transmission of nerve impulses in the brain, so that despite its rich protein content, *L. sativus* cannot be used as a food source. In all the above cases (leafy vegetables, sunflower and *L sativus)*, transfer of single gene for oxalate decarboxylase to produce transgenic plants will improve these plant species by reducing the oxalate content. This new gene will also allow the development of a 'diagnostic kit to measure oxalate in blood and urine', since prevalent methods for clinical tests are quite expensive.

Gene for a Seed-Specific Protein with Nutritionally Balanced Composition of Amino Acids : Prof. Datta's laboratory also reported, for the first time (in 1992) isolation of a gene from *(BC-45) Amaranthus* that encodes a seed-specific protein rich in all essential amino acids including lysine and sulphur rich amino acids. Interestingly, the amino acid composition of this protein corresponds to that of WHO recommended protein standard for the optimum human nutrition. It is known that seed proteins in cereals as well as in legumes (peas, etc.) are! deficient, the cereal proteins being deficient in lysine and pulses being deficient in sulphur containing amino acids. In view of this, the gene isolated from *Amaranthus* will be utilized in future for compensating the amino acid deficiencies of many seed proteins, once it is genetically engineered into the target crop plants. Efforts have been made jointly by Prof. Asis Datta and the Department of Biotechnology, Govt. of India, to seek patent rights for this gene in USA and Europe. The gene has also been successfully used for production of transgenic potato.

Sequencing of a Gene or a DNA Fragment

Once a gene or a DNA fragment has been cloned, its further study involves DNA sequencing. The technique of DNA sequencing was very laborious till 1975, when a breakthrough was made in DNA sequencing methods. The first two of the different methods described below are now routinely used for determination of DNA sequences although several other methods have been proposed for this purpose.

Chemical Degradation Method of Maxam and Gilbert : In this method, which is illustrated in Figure, following steps are involved: (i) Label the 3' end of DNA with ^{32}P. (ii) Separate two strands, both labelled at 3' ends, (iii) Divide the mixture in four samples, each treated with a different reagent having the property of destroying either only G, or only C, or G and A or C and T. The concentration of reagent is so adjusted that 50% of target base is destroyed, so that fragments of different sizes having ^{32}P are produced, (iv) Electrophorese each of the four samples in four different lanes of the gel.

(v) Autoradiograph the gel and determine the sequence from positions of bands in four lanes as shown in Figure.

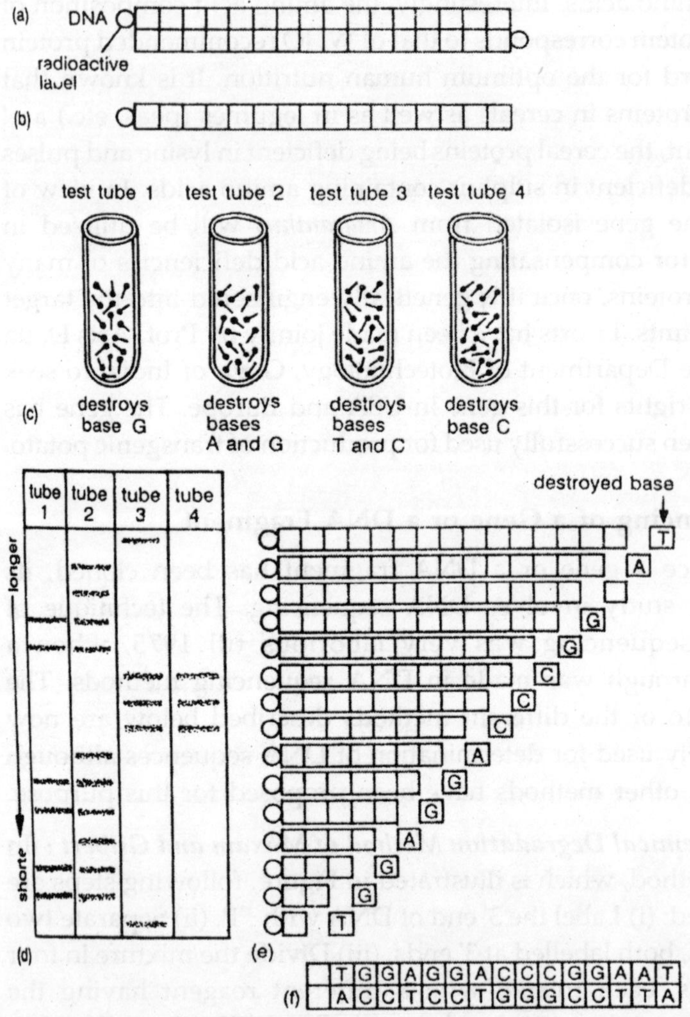

Different steps involved in Maxam and Gilbert's method of DNA sequencing.

Dideoxynucleotide Synthetic Method of Sanger : Frederic Sanger (who won Nobel Frize twice) had initially developed a method for DNA sequencing, which utilized DNA polymerase to extend DNA chain length. This was termed plus-minus method. Subsequently, he developed a more powerful method, utilizing single stranded DNA as template for DNA synthesis, in which 2,3, dideoxynucleotides were incorporated leading to termination of DNA synthesis.

These dideoxynucleotides are used as triphosphates (ddNTP) and can be incorporated in a growing chain, but they terminate synthesis (Fig.), since they cannot form a phosphodiester bond with next incoming deoxynucleotide triphosphate (dNTP). Following steps are involved in Sanger's dideoxy method for DNA sequencing, (i) Four reaction tubes are set up, each containing single stranded DNA sample (cloned in M13 phage) to be sequenced, all the four dNTPs, an oligonucleotide sequencing primer (radioactively labelled) and an enzyme for DNA synthesis (DNA polymerase I = sequences). Each tube also contains a small amount (much smaller amount relative to four dNTPs) of one of the four ddNTP so that four tubes have each a different ddNTP, bringing about termination at a specific base-adenine (A), cytosine (C), guanine (G) and thymine (T). (ii) The fragments, generated by random incorporation of ddNTP leading to termination of reaction, are then separated by electrophoresis on a high resolution polyacrylamide gel. This is done for all the four reaction mixtures on adjoining lanes in the gel. (iii) The gel is used for autoradiography so that the position of different bands in each lane can be visualized, (iv) The bands on autoradiogram can be used for getting the DNA sequence.

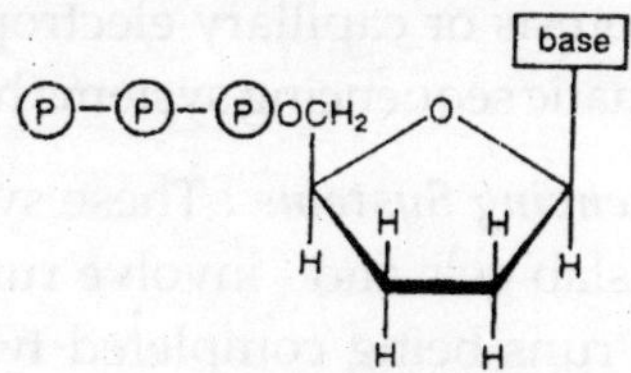

A dideoxy base analogue, showing dideoxy sugar (absence of oxygen at 2′ and 3′ positions).

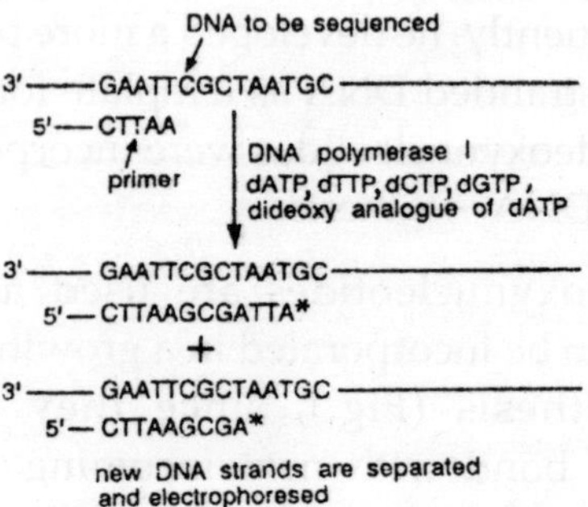

Strategy used in chain termination using dideoxy analogue of a base (only reaction with ddATP is shown).

DNA Sequencers of Automatic Nature

(Slab-gel Versus Capillary Sequencers)

A variant of the above dideoxy-method was later developed, which allowed the production of automatic sequencers (.e.g. Perkin Elmer's ABI Prism 310 or 3700 Genetic Analyser; Amersham Pharmacia Biotech's MegaBACE 1000 and 4000). In this new approach, different fluorescent dyes are tagged either to the oligonucleotide primer (dye primers) in each of the four reaction tubes (blue for A, red for C, etc.), or to each of the four ddNTPs (dye terminators) used in a single reaction tube; when four tubes are used, they are pooled. After the PCR reaction is over, the reaction mixture is subjected to separation of synthesized fragments through electrophoresis (Fig.). Depending upon the electrophoretic system used, whether slab gel electrophoresis or capillary electrophoresis, following two types of automatic sequencing systems have been designed.

Slab Gel Sequencing Systems : These systems make use of ultrathin (75 (im) slab gels and ' involve running of at least 96 lanes per gel, the runs being completed twice daily. In these systems, automation in sample loading of sequencing gels has also been achieved, by using a plexiglass block having wells that are same distance apart as the comb teeth cut in a porous membrane that is used as a comb for drawing samples by capillary action. Each well in plexiglass block is filled with a sample (PCR dideoxy-reaction mixture), so that when the porous membrane comb is lowered onto the sample wells in the plexiglass block, the samples are drawn up automatically into the comb teeth by capillary action. Using this approach of employing porous combs, automated loading of upto 192, 384 or "480 samples per gel has been achieved. The porous comb with the samples is placed between the glass plates of the gel apparatus above the flat surface of the polymerized gel and the samples are driven from the comb into the gel by electrophoresis.

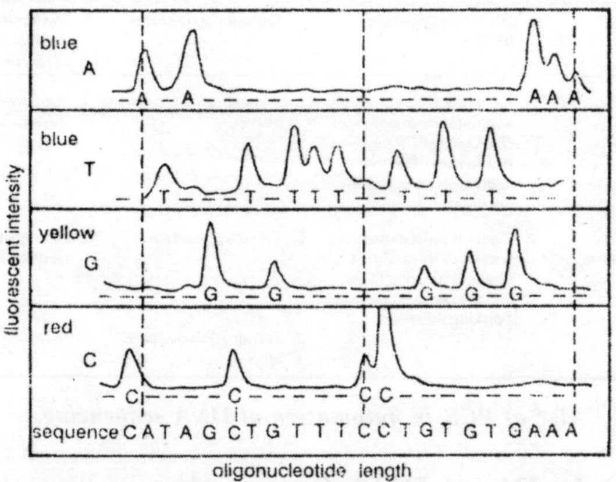

Fluorescence detection of oligonucleotide fragments produced by Sanger's dideoxy method (modified from Nature 321 : p 674: 1986).

Capillary Gel Sequencing Systems : In these systems, slab gel electrophoresis is replaced by capillary gel electrophoresis to analyse DNA samples. In these systems, instead of scanning DNA as it migrates through 96 lanes each in a series of poured slab gels, DNA is sent through a set of 96 capillary tubes filled with a polymerised gel. As the DNA fragments pass through these capillary tubes, DNA fragments are scanned. In the original models of the above old slab gel machines, gels must be poured and reagents frequently reloaded, interrupting the sequencing. In capillary gel sequencing systems, on the other hand, the robot moves the DNA samples and reagents through the tubes continuously, requiring attention only once a day. The system produces a steady flow of data, each signal representing one of the four DNA bases (adenine, cytosine, guanine and thymine). More recently (in 2002), 'Amersham Pharmacia Biotech' has marketed its improved model MegaBACE 4000, which makes use of 384 capillaries thus increasing manyfold the speed of sequencing. As mentioned above, due to competition with these capillary sequencing systems, automation has now been achieved in several models of slab gel systems also.

-Genome DNA *-Cloned DNA*	*DNA Amplification (PCR)*	*Sequencing reactions*	*Automated analysis of sequencing Reactions*
1. Mutation screening for known genes	1. Generation of sequencing templates from genomic DNA using PCR (ligation of adapters of known sequences for designing primers)	1. Annealing of sequencing primer.	1. Electrophoretic separation of sequencing reactions.
2. Isolation of DNA fragments by cloning	2. Generation of sequencing templates from cloned fragements using PCR and primers to vector sites flanking the insert.	2. Labelling reaction.	2. Computer registration of sequencing data.
		3. Termination reaction. 4. Stop.	

Use of PCR in automation of DNA sequencing.

PCR's use in Direct DNA Sequencing

(Also Called Ligation Mediated PCR = LMPCR)

Polymerase chain reaction (PCR) has also been used for sequencing the amplified DNA product. This method of DNA sequencing is faster and more reliable and can utilize either the whole genomic DNA or cloned fragments for sequencing a particular DNA segment. The DNA sequencing using PCR involves two steps : (i) generation of sequencing templates (double-stranded or single stranded) using PCR and (ii) sequencing of PCR products either with the thermolabile DNA polymerase or with the thermostable Taq DNA polymerase. Thus the DNA sequencing method using PCR eliminates the need of cloning the DNA in single-stranded DNA phage vector i.e. M13.

Although double-stranded DNA product of PCR can be utilized for sequencing, this may cause difficulty due to reassociation after denaturation thus preventing the sequencing primer from annealing to its complementary sequence to allow extension. To reduce this problem, either a variant of the standard method for sequencing double-stranded DNA is employed or single-stranded DNA templates are produced following asymmetric PCR (Consult author's book *"Biotechnology and Genomics"* for PCR).

A number of thermolabile DNA polymerases have been used for sequencing of *in vitro* amplified DNA. Alternatively, Taq polymer (thermostable enzyme) used for PCR, can be used for sequencing reaction. In either case Sanger's synthetic method involving incorporation of dideoxynucleotide (ddNTP) of chain termination, is used for sequencing. However, Maxam and Gilbert's chemical degradation method can also be used.

In Sanger's method, as usual, four mixtures are prepared, each using one of the four ddNTP. The sequencing primer is labelled with ^{32}P and the mixtures with amplified DNA, Taq polymerase and appropriate buffer are incubated at 70°C for 5 min. The reaction is stopped addition of polyacrylamide stop solution in each and mixtures are run on polyacrylamide sequencing gel to obtain the ladders, which can be read by computer or manually. This method using PCR helped in automation of DM sequencing (Fig.).

Transcription Method for DNA Sequencing

Recently, DNA sequencing was successfully achieved using RNA polymerase enzyme for transcription, instead of using 'sequenase' or 'thermo-sequenase' used in Sanger's dideoxynucleotide method of chain termination. This method of sequencing described 'transcriptional sequencing' makes use 3′dNTPs for terminating transcription in the same manner as ddNTPs are used for terminating DNA synthesis in Sanger's method (whenever 3′dNTP is available at 3' position, no OH or PO_4 group will be available for the formation of phosphodiester bond). The 3′dNTP can be labeled with fluorescent (rhodamine) dyes, so that four dNTPs with four colour dyes (TMR-3′dUTP, ROX-3′dCTP, R6G-3′dATP, R110-3′dGTP) can be used along with sufficient quantities of all the four NTPs (ribonucleoside triphosphates). This will allow RNA synthesis to proceed normally with the help of RNAP, till occasionally a 3′dNTP (dye labelled) is incorporated. This will yield RNA chains of different lengths, each having at its terminus, a specific dNTP, which will be identified by its colour on an automatic sequencer. The method is believed to be superior, firstly, because incorporation of NTP during transcription does not require a primer, thus making it faster (240 bases/sec) than DNA replication by Taq polymerase (60 bases/sec), *secondly,* because a large amount of product can be obtained from small template (600 molecules of RNA from one DNA molecule), and *thirdly,* because direct PCR sequencing can be done, without the elimination of primers and, 2'-dNTPs. The template in this method will have to be prepared by PCR using primers, to which phage promoter sequences are appended, which facilitate binding of RNAP for transcription. The residual primers and 2'-dNTPs need not be removed, because they will not interfere with RNA synthesis during transcription.

DNA Sequencing by Hybridization Employing Microarrays on DNA Chips : The main objective of several genome projects is the large-scale sequence analysis of genomes and its application to individual case studies in medicine and research. Sequencing by hybridization for detection of single nucleotide polymorphisms (SNPs) using DNA microchips was developed to meet this objective. A sequencing microchip is developed either by immobilizing or by parallel synthesis of a complete set of 8-mers (i.e. $4^8 = 65536$) on a solid support. The dimension of the microchip can be further reduced by using printing technology and robotics. The technique involves hybridization of a short DNA fragment of unknown sequence with a set of short oligonucleotides, particularly 8-mers of known sequences. Identification and analysis of a set of overlapping oligomers that form perfect duplexes with the DNA of interest permits reconstruction of the target DNA sequence. For example, if we wish to determine the sequence of a 17 base target DNA fragment, and hybridize it with complementary, overlapping 8-mers to form perfect duplexes, the overlap of seven or fewer adjacent 8-mers showing perfect match enables reconstruction of the complementary DNA sequences of the fragment.

Oligonucleotides shorter or longer than the 8-mers can also be used for sequencing, but while the use of shorter oligos leads to decrease in stability of the duplexes formed, the use of longer oligos leads to loss of discrimination between perfect base pair matches and mismatches. Therefore, the sequencing microchip is mainly composed of 8-mers. The hybridization is generally carried out at lower temperature, so that discrimination between perfect and imperfect duplexes can be achieved.

The limitation of the above technique is that the fragment to be sequenced should be small in size, although it should still be longer than n-1, where n is the length of the oligonucleotides immobilized on microchip. Analysis of long DNA fragments may be facilitated by fragmentation of the target DNA into smaller pieces prior to hybridization experiments. However, the method cannot be applied to repeat DNAs such as microsatellites or minisatellites.

***DNA Sequencing by using DE-MALDI-TOF Mass Spectrometry* :** Matrix assisted laser desorption/ionization mass spectrometry (MALDI-MS) has the potential to rapidly acquire DNA sequence information. It has been successfully used for confirmation of the sequences of short synthetic oligonucleotides by mass analysis of products resulting either from (i) synthesis failures (as in Sanger's method) or due to (ii) time-dependent exonuclease digestion. In both cases different products (oligonucleotides) will differ by single nucleotides. The mass of each fragment can be accurately determined and the fragments can be arranged by differences of single nucleotides, that can be identified by mass differences. This method of sequencing has been applied only to short oligonucleotides, since the resolution and sensitivity of MALDI-TOF goes down with increasing size.

For the purpose of DNA sequencing, MALDI-TOF mass spectrometry can also be used in combination with Sanger's termination reactions. In this approach, mass spectra of the four Sanger's sequencing reactions are overlaid and each sized product connected to one of the four bases. The advantage of this approach is that MALDI-TOF MS is quick, requiring only minutes per sample and no gels, radioactivity or fluorescent labelling is required. Although in theory this should work, in practice problems have been faced due to limited amounts of termination products and due to the presence of buffer salts and other components in the reaction mixture. The resolution was improved by introducing delayed ion extraction (DE-MALDI).

Gene Synthesis

There are two approaches available for the synthesis of genes : (i) When the detailed structure of a gene is available, this gene can be synthesized by a purely chemical method as done by H.G. Khorana for the synthesis of gene for a tRNA (reported for the first time in 1970). (ii) If the detailed nucleotide sequence of the gene is not available, one may utilize 'RNA-directed DNA polymerase' (reverse transcriptase) enzyme for the synthesis of the gene in question in the form of complementary DNA (cDNA) from the mRNA of the gene isolated in its pure form. This cDNA gene, however, would lack intron sequences and other gene sequences that are transcribed, but are removed during RNA processing. As discussed above, therefore, cDNA is often used for isolation of a gene either from (i) DNA extracted from living cells utilizing restriction digestion, electrophoresis and the technique of Southern blot hybridization, or from (ii) a genomic library.

tRNA genes' Chemical Synthesis : As outlined above, before one can start the chemical synthesis of a gene, the structure of gene should be known. The structure of a gene earlier could not be worked out by direct chemical analysis, since there were no means for isolating a gene (techniques for isolation of genes were developed later and were described earlier in this chapter). The structure of gene could, therefore, be inferred only from its product. For instance, if a gene is responsible for giving rise to a polypeptide chain and the structure of this chain is known, then from the genetic code dictionary, structure of the gene could be easily inferred. Such genes were initially considered to be too long to be synthesized, because an average gene contains about 1,500 base pairs. On the other hand, since tRNA molecules are fairly small in size (about 80 nucleotides), a gene responsible for giving rise to a tRNA molecule was within the reach for synthesis.

Synthesis of Gene for Yeast Alanyl tRNA : As readers may know, structures of large number of tRNAs are now known. The first tRNA whose structure could be known was yeast alanyl tRNA. R.W. Holley (who died in early 1993) and his coworkers (in 1965) gave the detailed structure of yeast alanyl tRNA. This information was immediately used by H.G. Khorana and his coworkers to deduce the structure of the gene for yeast alanyl tRNA. The structure of this gene would obviously be such that one of the two strands of DNA (gene) would be complementary to base sequence of yeast alanyl tRNA. The other strand would then automatically have the same sequence as in tRNA except that in place of uracil of tRNA, there would be thymine in DNA.

Synthesis of such a long chain (77 base pairs) of double-stranded DNA was rather difficult in 1965 (not in 1990's and thereafter) but Khorana and his coworkers had extensive experience of synthesizing DNA of known base sequences. It was decided that such a long chain could not be synthesized by adding a single base each time. Therefore, it was decided that small oligodeoxyribonucleotides ranging in length from 5 to 20 nucleotides should first be synthesized. These segments would be single stranded and would cover the whole length of both the strands of DNA. These would then be joined to form double-stranded DNA, 77 nucleotide pairs long. In actual synthesis of yeast alanyl tRNA gene, the following steps were involved.

(a) *Synthesis of Oligonucleotides* : Fifteen oligonucleotides ranging from pentanucleotide (5 bases) to an icosanucleotide (20 bases) were synthesized. Synthesis was conducted through condensation between hydroxyl group at the 3' position of one nucleotide and phosphate group at 5' position of the second nucleotide. In order to bring about condensation, all other functional groups, not taking part in condensation were protected using specific protective groups. After protecting the groups, reaction between a nucleotide with protected 5' end and another nucleotide with protected 3' end proceed according to Figure. Subsequently, condensation was also brought about between groups of two, three or four nucleotides.

(b) *Synthesis of Three Duplex Fragments* : With the help of 15 oligonucleotides (single stranded), three duplex fragments, each with a single stranded end, were prepared (Fig.).

These three fragments consisted of the following :

(i) the first 20 nucleotides with the nucleotide residues 17-20 as single stranded; this fragment was called 'A';

(ii) the nucleotide residues 17-50 with single-stranded regions 17-20 and 46-50; this segment was called 'B'; and

(iii) the nucleotide residues 46-77 with single- stranded region 46-50; this segment was called 'C.

(c) Synthesis of Gene from Three Duplex Fragments : The three fragments 'A', 'B' and 'C, were joined to give complete gene in each of the two following ways :

(i) In one scheme, 'A' was joined to 'B' taking advantage of the overlap in residues 17-20; 'C was then added with the overlap in the region 46-50 residues. The complete double-stranded DNA with 77 base pairs representing the gene was thus ready,

C–An A–Bz

NC CH_2CH_2O — P — O 3' 5' OH (1) + HO — P — O 3' 5' OAc (2)

mesitylene sulphonyl chloride

C–An A–Bz

NC CH_2CH_2O — P — O 3' 5' O — P — O 3' 5' OAc

Condensation between two nucleotides with protected 5' OH and 3' OH groups in the sugars and protected amino groups in the nitrogenous bases (Ac = acetyl, An = anisoyl and Bz = benzoyl groups).

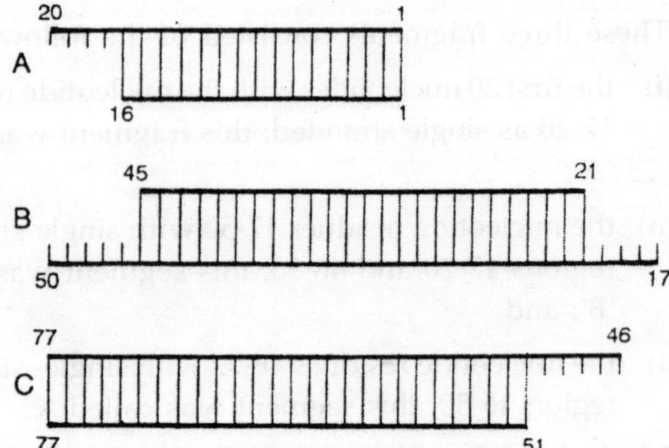

Three duplex fragments, with single-stranded sticky ends, produced for the synthesis of yeast alanyl tRNA gene by H.G. Khorana.

(ii) In the other scheme, 'B' was first added to 'C and to this 'A' was added in the end to obtain the complete gene. This gene, which was synthesized in 1970, was hopefully used for subsequent work, since this gene could replicate and make its own copies.

Synthesis of Gene for a True Precursor tRNA **:** Before Khorana could complete the synthesis of gene for yeast alanyl tRNA in 1970, as outlined above, it became obvious that tRNA was not the first direct product of transcription. Instead a precursor molecule is first synthesized which subsequently, after losing segments of RNA by cleavage, give rise to tRNA. Obviously, therefore, the actual gene for yeast alanyl tRNA will be longer than the DNA duplex synthesized by Khorana. In view of this, Khorana subsequently initiated synthesis of a gene for *E. coli* tyrosine suppressor tRNA precursor.

DNA duplex which will give rise to this tRNA precursor, was synthesized in the form of 26 small oligonucleotide segments. These were then arranged into six DNA duplex fragments having single-stranded ends. These six fragments gave rise to presumed gene for *E. coli* tyrosine suppressor tRNA precursor. This gene, however, still lacked promoter region and other sequences essential for processing.

Later, in 1979, Khorana reported completion of the total synthesis of a biologically functional tyrosine suppressor transfer RNA gene carrying all regulatory sequences. This gene was 207 base pairs long and included the following :

(i) a 51 base pairs long DNA promoter region;

(ii) a 126 base pairs long DNA corresponding to the precursor tRNA and

(iii) a 25 base pairs long duplex DNA corresponding to 16 base pairs adjoining CCA end of tRNA and the remainder, a modified sequence including *EcoRI* endonuclease specific sequence.

The details of the complete gene are shown in Figure. The complete synthetic gene was cloned in 'a vector by one of the gene cloning methods. On transformation of *E. coli*, the phage could multiply with the cloned gene.

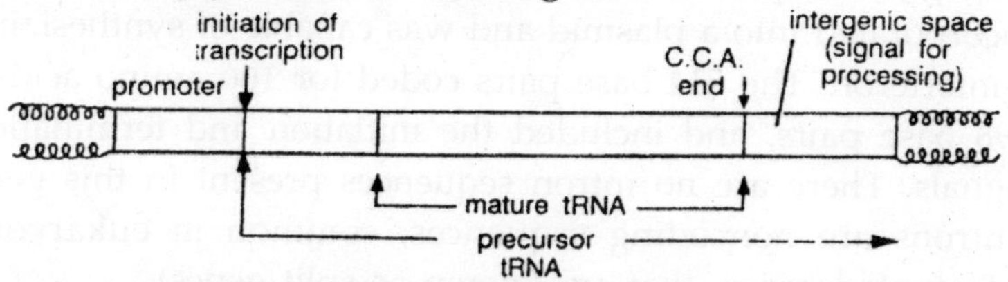

Different parts of complete E. Coli tyrosine suppressor tRNA gene (redrawn form Science, Vol. 203; 1979).

A Human Leukocyte Interferon Gene's Total Synthesis

Interferon is a protein, which has a property of inhibiting viral infection and cell proliferation and thus can be used for treatment of viral infection and malignancies (cancer). These properties have generated great interest in human interferons (IFNs). At least three kinds of interferon genes are known in human genome:

(i) leukocyte interferon genes (IFN-a genes),

(ii) fibroblast interferon genes (IFN-P genes), and

(iii) immune interferon gene (IFN-y genes).

Genes of different kinds of interferons differ and details of base sequences are now known for all of them.

In 1980, Weismann and coworkers published the nucleotide sequence of the first IFN-cc-gene derived from cloned IFN-cc cDNA. On the basis of this information, total synthesis of this human leukocyte interferon gene, 514 base pairs long, was achieved and published in 1981. Since there are 514 base pairs in the gene, it involved synthesis of DNA strands with 1028 nucleotides in a predetermined way. This was the longest gene synthesized till 1982.

The synthesis involved coupling of initial nucleotide onto a polyacrylamide resin to which further nucleotides could be added in pairs. Sixty-six (66) oligonucleotide segments varying in size from 14 to 21 bases were synthesized, which were then arranged in a predetermined way and joined. This gene was incorporated into a plasmid and was capable of synthesizing a-interferon. The 514 base pairs coded for 166 amino acids - 498 base pairs, and included the initiation and termination signals. There are no intron sequences present in this gene (introns are noncoding sequences, common in eukaryotic interrupted genes, that are known as split genes).

The plasmids with this newly-synthesized gene were transferred into bacteria *(E. coli)*, which are being utilized now for synthesis of interferon in industry. This has greatly reduced the cost of production of this drug (interferon) which was earlier sold at the rate of Rs. 16 million per 50 mg.

Machines for Gene Synthesis

Methods for synthesis of DNA molecules of known base sequences have improved dramatically during 1980's and 1990's. Therefore, genes can now be synthesized rapidly and in high yields. For instance a gene for a tRNA took 20 man years effort during 1965-70 (described earlier in this chapter), but the same gene can now be synthesized within a matter of few hours/days with the help of Gene Synthesis Machines developed in 1980's. The key innovations which made the availability of gene synthesis machines possible, include the following :

(i) development of silica based supports, which are insoluble and provide support for solid phase synthesis of DNA chains, and

(ii) development of stable protonated deoxyribonucleoside phosphoramidites (Fig.) as synthons, which are remarkably stable towards oxidation and hydrolysis and thus are ideal for DNA synthesis.

These two procedures can be used to prepare DNA molecules more than 100 nucleotides long within a few hours/ days. Different steps involved in solid phase synthesis of a DNA chain are shown in Figure . An outline of gene synthesis is shown in Figure.

Protection Chemistry and Amidites : The methods of gene synthesis make use of protection chemistry to minimize undesired side reactions during the synthesis. Two different types of protecting groups are needed : (i) permanent protection is needed for the primary amino groups of all heterocyclic bases and for the hydroxyl groups of the phosphate of all nucleotides; (ii) temporary protection is needed for 5' hydroxyl group of the incoming nucleotide.

No other functional groups need to be protected, since they do not react under the conditions used for gene synthesis. A fully protected nucleotide is called an amidite. A, C, G and T amidites are needed in all synthesizers. However, p-cyanoethyl phosphoramidites using the PAC protection scheme (see later for PAC) are used in Gene Assembler Synthesizers.

A protonated phosphoramidite (5′ hydroxyl is blocked by DMT = dimethoxytrityl group).

Different steps involved in solid phase synthesis of a DNA chain by the phosphite triester method.

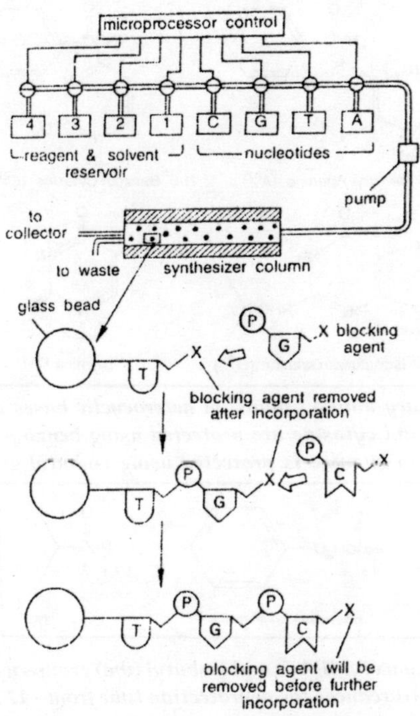

An outline of a gene synthesis machine, which synthesizes a gene automatically once the nucleotide sequences are entered in the microprocessor.

(a) Protecting Groups of Permanent Nature : These protecting groups are stable in all reaction steps and remain in position throughout the synthesis. They are removed only after the synthesis is completed. Primary amino groups on the heterocyclic bases are generally protected as amides (Fig.), using benzoyl group for A or C, and isobutyl group for G. Thymine has no amino group and thus needs no protection. Deprotection is done by using concentrated ammonia at 55°C overnight (~12 hours). However, if PAC (phenoxyacetyl) amidite protection strategy is used, the deprotection time can be reduced to 1 hour at 70°C. In this strategy, PAC group is used to protect A and G; isobutyl protection is used for C (Fig.). For protection of phosphate group in gene synthesizers, phosphite triester (phosphoramidite) chemistry is used. In this method, hydroxyl group on the phosphorus atom is protected using a methyl or P-cyanoethyl group (Fig.), the latter is

N-6 Benzoyl-Adenine (A^{bz})　N-6 Benzoyl-Cytosine (C^{bz})

N-2 Isobuturyl- Guanine (G^{ibu})　Thymine (T)

Protection of primary amino groups on heterocyclic bases as amides (amino groups in adenine and cytosine are protected using benzoyl group and amino group in guanine is protected using isobutyl group).

phenoxyacetyl (PAC)　isoburyl (IBU)

Structures of phenoxyacetyl (PAC) and isobutyl (ibu) groups used in PAC amidite protection strategy, (it reduces the deprotection time from - 12 hours to one hour).

NH—PAC

DMTr—O

$CNCH_2CH_2$—O

(an amidite fully protected)

Protection of - NH2 group by PAC and phosphate group by B cyanoethyl group to produce an amidite.

removed by treatment with concentrated ammonia after the synthesis is completed. This also means that the protecting groups in the bases as well as in the phosphate can be deprotected in one step using ammonia.

(b) Protecting Groups Temporary Nature : These protecting groups need to be removed after every step during the synthesis. The 4, 4' dimethoxytrityl group (DMTr) is used to protect the hydroxyl group at position 5' of the incoming nucleotide. It is removed in every step using mild acidic solutions (1% trichloracetic acid = TCA or 3% dichloracetic acid = DC A).

Oligonucleotide Support : The support used in oligonucleotide synthesis consists of two main parts : (i) the first nucleoside unit, from which the synthesis starts; (ii) the solid phase (an organic support to which the first nucleoside is linked by means of a spacer arm; Fig.). The solid phase technique helps in rapid and efficient washing of the growing oligonucleotide between the steps to remove excess reactants.

A solid support should have the following features : (i) should be insoluble in all solvents and reagents, (ii) should not swell under the conditions used for synthesis, (iii) should be inert towards chemicals used, (iv) should be macroporous (to give rapid access to reagents), (v) should have large surface area, (vi) should be rigid (to withstand solvent pressure applied during synthesis), (vii) should be capable of bearing functional groups (which can react with deoxyribonucleotides in a reproducible manner). The main four types of solid support include : (i) silica, (ii) glass beads (CPG), (iii) plastic (e.g. polystyrene) and (iv) cellulose paper. Of these four, glass needs to be derivatised to enable coupling of the first deoxyribonucleoside to it. An amino group is introduced for this purpose at the end of a spacer arm.

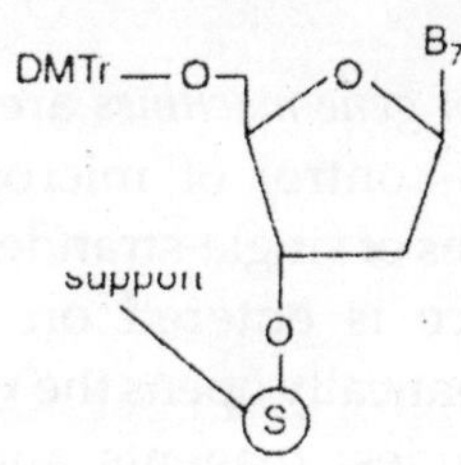

A nucleoside attached to a solid support (S).

The Synthesis Cycle : The synthesis of an oligonucleotide starts with the first deoxyribonucleotide (in its protected form) linked to the solid phase through its 3' end. All the four nucleotides are added as protected amidites. The reaction cycle then consists of the following four steps : (i) Deprotection (detritylation) : the solid support is washed with dichloroethane and the 5' hydroxyl group of deoxyribonucleotide is deprotected removing DMTr group (1.5 to 3 minutes of acid treatment); the reagents are removed by a wash with dichloroethane and anhydrous conditions for coupling reaction are achieved by an acetonitrile wash; (ii) Coupling : 3'-5' phosphodiester linkage is achieved using 3' phosphate group of the incoming protected deoxyribonucleotide (a phosphoramidite or phosphite triester); this is achieved by phosphite triester synthesis, in which phosphoramidite is first converted to a tetrazolide (using tetrazole as a weak acid); excess reagents are washed by acetonitrile. (iii) Capping : Any unreacted 5' hydroxyl groups anchored at the solid phase are protected by acetyl group using acetic anhydride (AC_2O + N-methylimidazole + 2,4,6-collidine in acetonitrile). This allows only the correct chains to elongate in subsequent coupling reaction; (iv) Oxidation : Before adding the next nucleotide, the phosphate triester is oxidized to the more stable 5-valent oxidation state, by treatment with a mixture of iodine, collidine and water in acetonitrile.

After the Synthesis : Following two steps are used to obtain the final product : (i) removal of permanent protecting groups and cleavage of oligonucleotide from the solid phase, (ii) analysis and/or purification of the product by chromatography or electrophoresis.

Several versions of *gene machines* are now available. These machines, under the control of microprocessor, synthesize specific short sequences of single-stranded DNA automatically. The desired sequence is entered on a keyboard and the microprocessor automatically opens the valves of the containers of successive nucleotides, reagents and solvents needed at each step, into a synthesizer column, which is packed, with tiny silica beads. These beads provide support on which DNA molecules are assembled.

Gene using PCR's Synthesis

The methods for gene synthesis utilized earlier, whether manually or through automatic synthesizers, involved the following steps: (i) synthesis of oligos, (ii) annealing of oligos to give duplexes with single-stranded cohesive or sticky ends and (iii) ligation of duplexes to obtain complete gene. This strategy was initially used by Khorana and was subsequently followed by many. Even while using automatic synthesizers, only the oligos are synthesized on the synthesizer, but the production of gene from these oligos is achieved later following the usual procedure. Crude synthetic oligonucleotides were sometimes also used for rapid generation of DNA fragments through PCR. In all these procedures oligos are first phosphorylated at the 5' ends, overlapping ends annealed, gaps filled by enzymatic extension at the 3'ends and finally nicks were joined with DNA ligase. The full length double-stranded DNA could be cloned in a plasmid/phage vector and multiplied in *E. coli.* Alternatively, it could be amplified by PCR, separated on gel, purified from gel and cloned. The above steps of phosphorylation, annealing, ligation and cloning could be dispensed with in some of the recent methods of gene synthesis. For instance, overlapping oligonucleotides could be extended through several rounds of PCR. Full length double-stranded DNA could then be amplified using short primers complementary to the 5'ends of two strands. In India, at NBRI, at Lucknow, Dr. Rakesh Tuli and his group standardized a

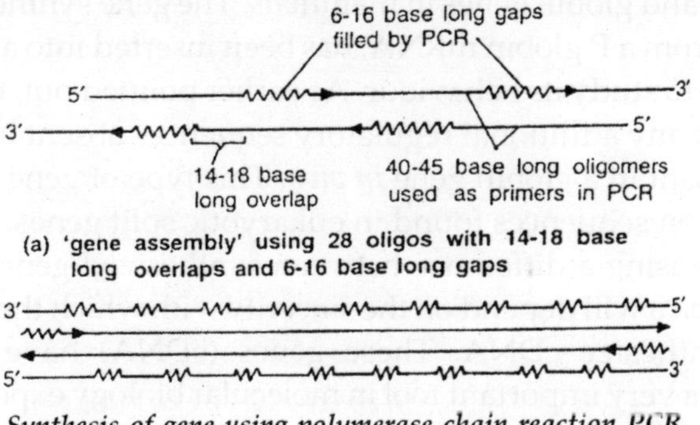

Synthesis of gene using polymerase chain reaction PCR.

method for construction of synthetic genes. In this method, overlapping oligos with gaps (6-16 bases long) between adjacent oligos were first synthesized on an automatic sequencer, the complementary oligos were annealed using 14-18 bases long overlaps and the assembled double- stranded DNA was used for PCR to give complete DNA sequence by several cycles of PCR (Fig.). Deep vent polymerase, instead of Taq polymerase was used to reduce error rate.

Synthesis of Genes from mRNA

H. Temin and D. Baltimore, in 1970 discovered the presence of 'RNA directed DNA polymerase' (reverse transcriptase) enzyme which has the ability of synthesizing DNA on RNA template. This enabled molecular biologists to synthesize complementary DNA (cDNA) using mRNA as template. If mRNA transcribed from a specific gene is made available in purified form, the complementary DNA (cDNA) synthesized with its help will represent the synthesized gene. To what extent this synthesized gene is a faithful copy of native gene will depend on the fidelity of copying mRNA and also on the stability of DNA thus synthesized. Moreover, since mRNA of a gene does not have the complete transcript of the gene *in vivo,* the synthesized gene will be smaller than the gene *in vivo.*

By copying eukaryotic purified mRNA, several genes have been artificially synthesized. Most important of these genes are the genes for sea urchin histone proteins, ovalbumin gene in chicken and globin genes in mammals. The gene synthesized as cDNA from a P globin mRNA, has been inserted into a plasmid in order to study its behaviour. As earlier pointed out, this gene will lack any additional regulatory sequences absent in mRNA but present in a globin gene *in vivo.* This type of gene will also lack intron sequences found in eukaryotic split genes. Another factor causing a difference between synthesized gene and the gene *in vivo* will depend on the sincerity with which the enzyme will synthesize cDNA. These genes (cDNA) have already become a very important tool in molecular biology experiments.

4

Gene Amplification

The techniques of recombinant DNA and gene cloning, which permit us to obtain an unlimited supply of identical copies of a gene sequence or DNA segment, which is cloned in a prokaryotic or a eukaryotic cell with the help of a vector. This technique, discovered around 1975, proved extremely useful in all experiments of molecular biology and genetic engineering and thus became an essential tool in all molecular biology laboratories. In 1985, yet another remarkable tool in molecular biology was discovered, which is known as polymerase chain reaction (PCR), later nick- named as 'people's choice reaction'. The methodology is so important that the prestigious journal Science published from USA considered PCR as the major scientific development of the year 1989, and had chosen Taq DNA polymerase, the enzyme used in PCR, as the molecule of the year 1989. Kary Mullis, who discovered this reaction in 1985, described in his article published in April 1990 issue of Scientific American, how he got the basic idea of this reaction, while driving through the red wood mountains of California, USA. In view of the significance that PCR and its varied applications achieved within a few years after its discovery, the 1993 Nobel Prize in Chemistry was shared by Karry Mullis for his discovery of PCR, with Michael Smith, for his discovery of site directed mutagenesis. The polymerase chain reaction is such a powerful technique that for majority of laboratory experiments, it has now largely replaced the gene cloning undertaken with the help of vectors. The basic principles and the methodology involved in PCR (including its various modified forms) and its various uses in molecular biology and biotechnology will be discussed in this chapter.

Fundamental PCR

In order to understand PCR, the readers should be familiar with the mechanism of DNA replication within the cell. DNA replication involves polymerization of nucleotides using a template DNA strand with the help of an enzyme DNA polymerase, but this reaction invariably requires a primer strand to which further nucleotides can be added using the DNA polymerase enzyme. If the primer strand is not available, the reaction cannot proceed. In the living cells this primer strand is not a DNA strand but is a small single-stranded RNA molecule synthesized with the help of RNA polymerase enzyme, which does not need a primer. In PCR, a similar reaction takes place in an 'eppendorf tube', where the primer strand is added from outside in the form of a deoxyribo-oligonucleotide, and DNA polymerase enzyme is added to

A thermal cycler for PCR amplification (the instrument has a microprocessor-controlled temperature cycling needed for PCR).

help in polymerization. Unlimited supply of amplified DNA is obtained by repeating the reaction, which is made possible by regular denaturation of freshly-synthesized double-stranded DNA molecules by heating it to 90-98°C. At this high temperature the two strands separate. Once the double-stranded DNA is made single stranded by heating upto 90-98°C, the mixture of these two strands with two primers that recognize

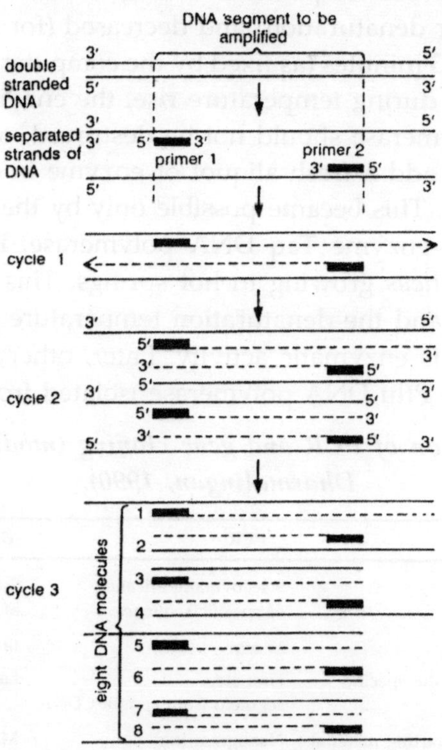

Basic reaction of the PCR (only three cycles of PCR are shown; in each cycle primers are shown by solid boxes, template strands are shown by continuous lines and newly synthesized strands are shown by broken lines). (Modified and redrawn from Dharmalingam. K. -Biospectra, Nov-Dec, 1990).

the sequences bordering the sequence to be amplified is cooled to 40-60°C. This allows the primers (which are in excess) to bind to their complementary strands through renaturation. The presence of Taq DNA polymerase enzyme and all the four essential nucleoside triphosphates allow synthesis of complementary strands in the usual manner. In a thermal cycler (Fig.), this process is automatically repeated 20-30 times (as predetermined by a computer device), so that in a single afternoon, a billion copies of the sequence flanked by the left and right primers can be produced (Fig.). In order to continue the synthesis, the temperature of the mixture is alternately increased (for denaturation) and decreased (for renaturation) once every 1-3 minutes (as fixed by the computer device). This requires that during temperature rise, the enzymatic activity of DNA polymerase should not be destroyed, otherwise one may have to add a fresh aliquot of enzyme in each cycle of amplification. This became possible only by the discovery of thermostable enzyme Taq DNA polymerase, isolated from *Thermus aquaticus* growing in hot springs. This enzyme acts best at 72°C and the denaturation temperature of 90°C does not destroy its enzymatic activity. Later, other thermostable enzymes like Pflu DNA polymerase isolated from *Pyrococcus*

Comparison of PCR and gene cloning (modified from Dharmalingan, 1990).

Parameter	*PCR*	*Gene cloning*
1. Final result	Selective amplification of specific sequence	Selective amplification of specific sequence
2. Manipulation	*In vitro*	*In vitro & in vivo*
3. Selectivity of the specific	First step segment from complex DNA	Last step
4. Quantity of starting material	Nanogram (ng)	Microgram *(fig)*
5. Biological reagents required	DNA polymerase (Taq polymerase)	Restriction enzymes, ligase, vector, bacteria
6. Automation	Yes	No
7. Labour intensive	No	Yes
8. Error probability	Less	More
9. Applications	More	Less
10. Cost	Less	More
11. User's skill	Not required	Required
12. Time for a typical experiment	Four hours	Two to four days

furiosus and Vent polymerase isolated from *Thermococcus litoralis*, were discovered and were found more efficient. These enzymes allowed automation of the entire process and automatic PCR thermal cyclers are now available (each for a price of around Rs 2-4 lakhs), which can amplify DNA sequences at a fast speed unattended.

As pointed out earlier, PCR may eventually replace gene cloning technique. This technology is much easier and requires much smaller quantity (nanogram or ng = 10^{-9}g) of DNA. For cloning, DNA is needed, as the starting material, in microgram (μg) quantities. A comparison between PCR and gene cloning is given in Table.

Different Schemes of PCR

We have already discussed the basic outline of PCR technique, outlining how billions of copies of a DNA sequence can be obtained within a few hours, without the use of a vector and a host as done in gene cloning. However, there are several variations to this basic PCR technique, which allow varied applications of this technique.

PCR of Inverse Kind

In this technique, the amplification of those DNA sequences takes place, which are away from the primers and not those which are flanked by the primers. For instance, if the border sequences of a DNA segment are not known and those of a vector are known, then the sequence to be amplified may be cloned in the vector and border sequences of vector may be used as primers in such a way that the polymerization proceeds in inverse direction i.e. away from the vector sequence flanked by the primers and towards the DNA sequence of inserted segment (Fig.). Similarly, if the gene sequence is known, one can use its border sequences as primers, for an inverse PCR reaction, to amplify the sequences flanking this gene e.g. the regulatory sequences (Fig.).

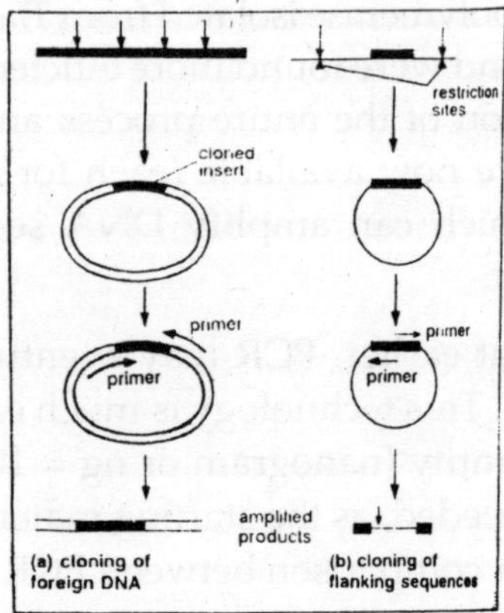

Inverse PCR : (a) amplification of genes using flanking vector sequences as primers; (b) amplification of sequences flanking the gene, using sequences at the two ends of the gene as primers. (Modified and redrawn from Dharmalingam, K.-Biospectra, Nov.-Dec, 1990).

PCR of Anchored Type

In the basic PCR technique and the inverse PCR, one has to use two primers representing the sequences lying at the ends of sequences to be amplified. But sometimes, we may have knowledge about sequence at only one of the two ends of the DNA sequence to be amplified. In such cases, anchored PCR may be used, which will utilize only one primer instead of two primers. In this technique, due to the use of one primer, only one strand will be copied first, after which a poly-G tail will be attached at the end of the newly synthesized strand. This newly synthesized strand with poly G tail at its 3' end will then become template for the daughter strand synthesis utilizing an anchor primer with which a poly-C sequence is linked to complement with poly-G of the template. In the next cycle, both the original primer and anchored primer will be used for gene amplification (Fig.).

PCR for Site Directed Mutagenesis

This technique is used for introducing mutations at the desired place in a DNA sequence by altering the sequences of primers (Fig.). Since mutations are introduced only through primers, mutations are limited to the ends of the gene sequence. A variation of this technique allows mutations to be introduced at any place of interest in the gene. The method is described as overlap extension, which works as follows

In two separate PCR reactions, a particular gene is amplified into two separate segments. In each of the two reactions, there is one primer at the end of the gene and the other (with desired alteration for mutation) internal to the sequence. The internal modified primers in two reactions are complementary to one another, so that the amplified products will have their ends internal to the original sequence. These internal ends of products in two reactions will overlap, so that if the two PCR products are mixed and used again for PCR, their internal ends function as primers. Extension of these primers by Taq polymerase results in the formation of a complete gene, with mutations incorporated at internal sites (Fig.).

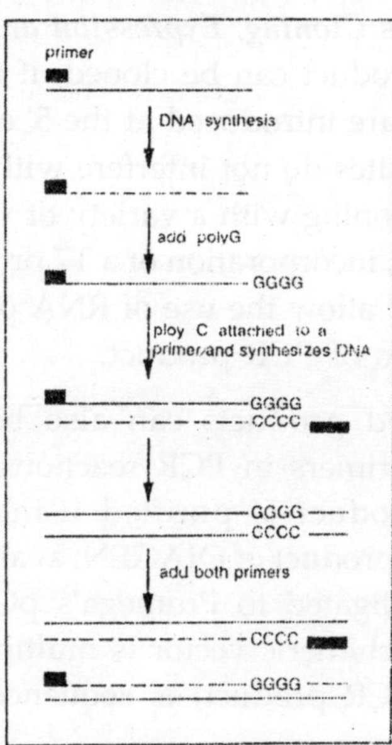

Anchored PCR, using a single primer in the first cycle (see text for details). (Modified and redrawn from Dharmalingam, 1990).

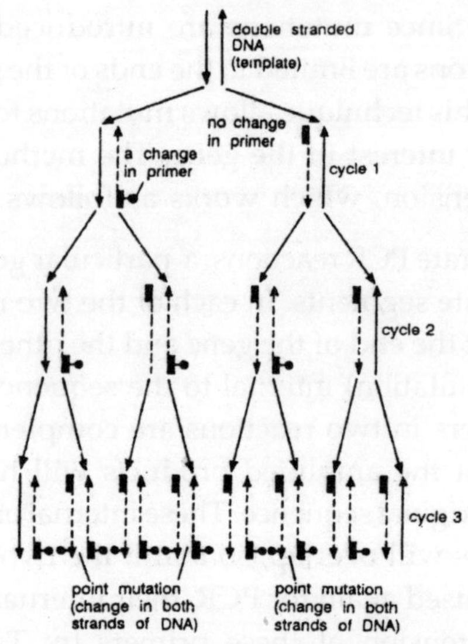

Site directed mutagenesis using PCR (minor changes are introduced in primers; these minor changes do not interfere with PCR). (Modified and redrawn from Dharmalingam, K.-Biospectra, Nov.-Dec, 1990).

PCR Product's Cloning, Expression and Sequencing : The amplified PCR product can be cloned, if restriction sites for specific enzymes are introduced at the 5' end of each primer. These restriction sites do not interfere with amplification and can be used for cloning with a variety of vectors in the usual manner. Similarly, incorporation of a T7 promoter at the 5' end of one primer will allow the use of RNA polymerase enzyme to get RNA copies of PCR product.

PCR amplified products can also be cloned, without using modified primers in PCR reactions. For this purpose the amplified product is purified using QIA quick PCR purification kit (a product of QIAGEN, available from Genetix, New Delhi) and ligated to Promega's pGEM-T or pGEM-T Easy vector. The chimeric vector is multiplied in *E. coli* cells and the insert (PCR product) is sequenced using universal M13. primers.

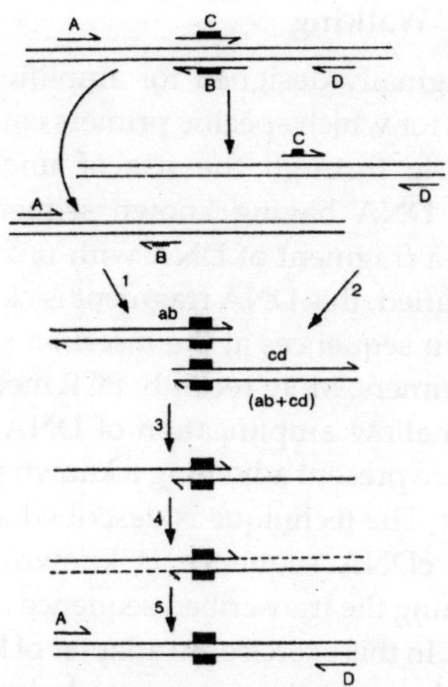

Introduction of a mutation at an internal site using overlap extension (two pairs of primers A & B and C & D are used. B and C represent internal sites and overlap each other). (Modified and redrawn from Dharmalingam, K.-Biospectra, Nov.-Dec, 1990).

PCR with Asymmetric Feature for DNA Sequencing : PCR products can also be used as templates for direct DNA sequencing (ligation mediated PCR = LMPCR). Single-stranded DNA can be produced for this purpose using asymmetric PCR, in which the two primers are used in 100 : 1 ratio, so that after 20-35 cycles of amplification, one primer is exhausted and single- stranded DNA is produced in the next 5-10 cycles (Fig.). For more details about PCR assisted DNA sequencing.

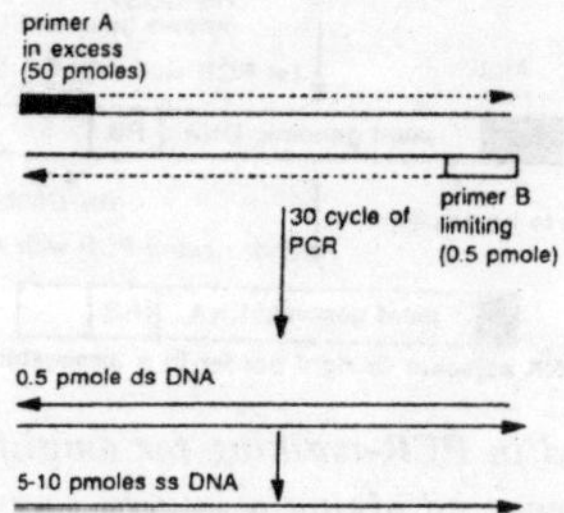

Asymmetric PCR for generating single stranded DNA for DNA sequencing.

Process of PCR Walking

PCR was originally designed for amplification of known DNA sequences, for which specific primers could be designed. This was possible through the use of uncloned DNA, if amplification of DNA having known sequences is desired. However, when a fragment of DNA with unknown sequence needs to be amplified, this DNA fragment is cloned in a vector, so that the known sequences at the insertion site of the vector can be used as primers. More recently PCR methods have been designed, which allow amplification of DNA with unknown sequences, that are present adjoining a known sequence within a DNA fragment. The technique is described as PCR walking. For instance, if cDNA sequence is known, the regulatory sequences adjoining the transcribed sequence in genomic DNA can be amplified. In this exercise, an adapter of known sequence is first ligated to the restriction fragments derived from genomic DNA, that contains the unknown sequence to be amplified. One of the two primers is designed from the known gene sequence (described as gene-specific primer) and the other primer is designed on the basis of the adapter of known sequence, that is ligated to the restriction fragments of genomic DNA. Consequently, PCR walking involves the following three

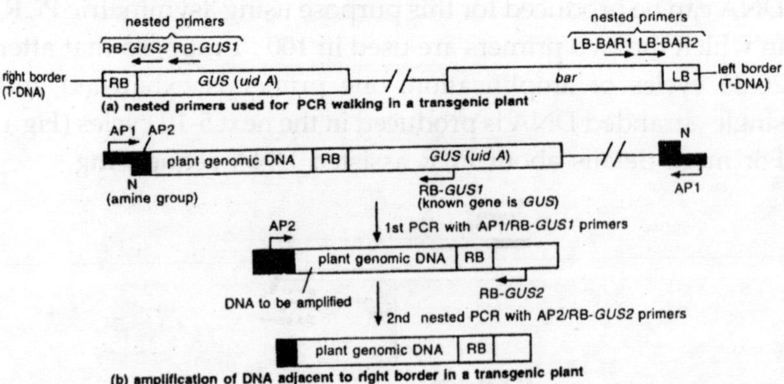

Different steps involved in PCR-walking for amplification of an unknown sequence adjoining a known sequence.

steps (Fig.) : (i) restriction enzyme digestion to produce a large number of genomic DNA fragments; (ii) ligation of adapters to restriction fragments and (iii) PCR amplification using one adapter primer and the other gene specific primer.

The adapter for the restriction fragments for PCR walking is designed to have an amine group on the 3' end of the lower smaller strand, to disallow extension of this end of the adapter, thus preventing generation of AP1 primer binding site (this is a feature of a method described as 'vectorette PCR'). Further, the adapter primer is shorter than the adapter itself and binds to the end of adapter , ligated fragments. The adapter also has inverted terminal repeats, so that if the PCR product contains double- stranded adapter sequence at both ends (due to non-specific DNA synthesis), the ends of individual DNA strands will form 'panhandle' structures following every denaturation step; this suppresses their further amplification, thus permitting

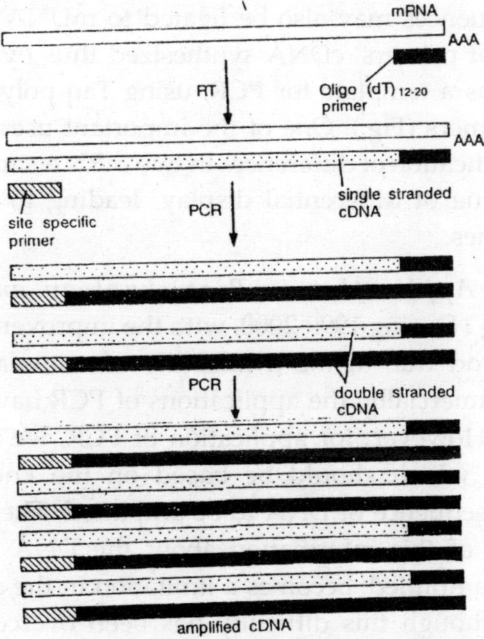

Different steps involved in RT-PCR (reverse transcribed-PCR) to obtain cDNA from mRNA.

amplification of only those regions, which are adjacent to gene-specific sequence, used for designing the second primer (this is called 'suppressor PCR'). The technique has been used for PCR walking upstream from exon of the human tissue-type plasminogen activator (tPA) gene, and upstream of the human transferrin receptor gene.

Method of RT (Reverse Transcribed)-PCR : In a number of situations, when mRNA transcript needs to be examined, reverse transcription is first used to obtain cDNA, which is then used as a template for PCR reaction. For this purpose, mRNA is first isolated, purified and subjected to reverse transcription with the help of reverse transcriptase enzyme. Often the synthesis of cDNA is initiated by primers that are either complementary to poly A tail (for mRNA) or complementary to specific sequences known to be present on one or more specific subsets of RNA species. Adapters of known sequences may also be ligated to mUNA to facilitate designing of primers. cDNA synthesized thus by RT-PCR, is then used as a template for PCR, using Taq polymerase and specific primers (Fig.). One of the important uses of RT-PCR is the identification of differences in two mRNA samples, using the technique of differential display, leading to isolation of specific genes.

PCR's Application in Biotechnology and Genetic Engineering : During 1990-2000, with the improvement of PCR protocols and due to the availability of automatic thermal cyclers commercially, the applications of PCR have increased manyfold. However, for application of PCR, we need a pair of primers which should be based on the knowledge of nucleotide sequence of DNA to be amplified. Therefore, non-availability of this information about the DNA segment or gene to be amplified, becomes a limitation in the applications of PCR, although this difficulty has been overcome for the study of DNA polymorphism, through the use of random

DNA primers. Some of the applications of PCR technology, while describing molecular markers.

Gene Tagging Employing PCR

PCR has also been utilized for developing molecular markers closely linked to specific genes of economic importance. For instance in tomato, 144 random primers were used to produce 625 PCR products from a set of near isogenic lines (NILs which differ only for the presence and absence of a specific gene) for *Pseudomonas* resistance. At least three of these PCR products were present in one and not in the other of a pair of NILs showing linkage with *Pseudomonas* resistance gene *(Pto)*. These PCR products could be mapped close to *Pto*. Such tagging of genes through PCR will be used in future for maker aided selection (MAS) in plant breeding (selection of desirable plants in segregating populations) and also for isolation of genes.

PCR's Function of Confirming the Transferred Gene's Presence : When a gene is transferred with a vector to cultured cells or organisms, primers can be designed to conduct PCR for amplification of the gene sequence so transferred. This technique allows confirmation of the transfer and maintenance of the gene of interest. In gene therapy experiments, the transfer and presence of a marker gene for neomycin resistance *(NeoR)* could be detected in the blood of patients, even after 60 days, although the frequency of marked lymphocytes gradually decreased necessitating a second transfusion of such marked cells. Similarly, ADA (adenosine deaminase deficiency) gene was transferred for therapeutic purposes and PCR was used to detect the presence of this gene upto 6.5 months, so that the patient (4-year-old girl) was then placed on a programme of maintenance infusions at 3-5 months intervals. Thus PCR proves to be of immerse help in monitoring a gene in genetic engineering or gene therapy experiments.

PCR's use Human Genetics

PCR has also found extensive use in human genetics. Some of these uses include the following.

Prenatal Diagnosis using PCR : Prenatal diagnosis of sickle cell anaemia with enhanced sensitivity was perhaps the first application of PCR for diagnosis of a human disease. Using PCR, this is done in less than one day in contrast to several days/weeks needed, when southern blots are used for hybridization with a probe. The test has also been used for diagnosis of phenylketonuria (PKU), (3- thalassemia, hemophilia, etc. The PCR product in all these cases is examined using a labelled probe, to suggest whether or not mutant sequence causing the disease is found. Sometimes, RFLP pattern of PCR products in healthy and defective feti (singular = fetus) differ, thus enabling prenatal diagnosis. In still other cases, PCR product may be sequenced to reveal the difference.

Recombining Data Employing PCR : PCR can be used for DNA amplification, utilizing individual sperms, which are the products of meiosis and recombination. Therefore, by studying PCR products from a large number of individual sperms (pollens in case of plants), one can calculate the proportion of sperms that are recombinant for linked markers. This allows construction of genetic maps at a resolution which is not possible with pedigree analysis. Microdissected chromosome bands are also used for PCR to allow physical location of genetic lesions.

PCR's Use in Sexing of Embryos : Since DNA sequences in single cells can be studied using PCR, sex of human or livestock embryos, fertilized *in vitro*, can be determined before implantation. For this purpose, PCR primers and probes specific for sex chromosomes (e.g. ZFY or ZFX sequences) can be used. Y-specific primers and probes in humans are available for this purpose. This technique can also be used to detect sex-linked disorders in the fertilized embryos.

PCR use in DNA Fingerprinting

Some well-characterized sequences of microsatellites are being used for designing primers, so that DNA fingerprinting may be achieved through PCR. PCR allows amplification of DNA from individual hairs, stains of blood or seminal fluid having partially degraded DNA, which could not be used earlier' for characterization of individuals.

What is Electronic-PCR (ePCR)? : In the post-genomic era, massive data on DNA sequences are now available in different databases. The approach of data mining using bioinformatics tools is employed for making good use of data. This data can also be utilized for conducting PCR on the computer. For instance, primers for a specific-conserved DNA sequence can be designed and used on the computer to find out the expected length of the amplified product. This approach has been successfully utilized for conducting electronic PCR (ePCR) without the use of PCR machine or PCR reaction mixtures.

5

Gene Cloning

This is achieved through construction of chimeric DNA molecules for the purpose of obtaining molecular probes and gene libraries. These different uses of vectors and the methods of cloning will be briefly described in this chapter. A brief account on restriction enzymes is also included in this chapter, since these enzymes are used regularly as a routine in all DNA cloning and recombinant DNA experiments.

Restriction Enzymes for Cloning

For cloning of DNA, often we need to cut the DNA at specific sites, which are recognized and cleaved by specific enzymes (endonucleases, which cleave DNA at internal sites), described as restriction enzymes. Also by locating the positions of cleavage sites of a number of restriction enzymes, restriction maps can be prepared (see later in this chapter). These restriction enzymes recognize short sequences of double stranded DNA as targets for cleavage.

Different enzymes recognize different, but specific

sequences, each ranging in length from 4 to 8 base pairs. Each enzyme is named by a three-letter (or four letter) abbreviation (letters are italicized) that identifies its origin. Roman numerals (I, II, III, etc.) are added to distinguish several enzymes with same origin. For instance *EcoRl* is derived from *E. coli* and *Hpal* is derived from *Haemophilus parainfluenzae.*

Besides cleavage, modification in the form of methylation is also brought about by some enzymes called modification enzymes (sometimes also called methylases). This methylation distinguishes genes in different states of functioning. There are also enzymes which perform the function of both restriction (cleavage) and modification (methylation). Based on these attributes, restriction enzymes have been grouped into two classes :

(i) types II restriction enzyme systems (e.g. *EcoRl)* which have separate enzymes for restriction and modification and

(ii) type I *(EcoK* and *EcoB)* and type III (EcoPl, EcoP15) enzyme systems, in which same enzyme possesses both activities (bifunctional), although the restriction and modification sites differ in positions.

Of the above two classes of restriction enzymes, type II enzymes are most important for cloning purposes. The target sites for these enzymes are 4-8 bp with a symmetry (palindromes). Most type II restriction enzymes cleave DNA at unmethylated target sites. Some enzymes introduce staggered cuts, others generate blunt ends. Enzymes with 4 bp target sites are used when frequent cuts are desired and those with 8 bp (e.g. *Notl)* are used when rare cuts are desired to get long DNA segments. Otherwise majority of enzymes used have 6 bp target sites. Some of them can cleave both methylated as well as unmethylated targets, but majority of them cleave only unmethylated targets. Details of some of these enzymes are listed in Table.

Recognition sequences of several restriction endonucleases (an arrow in each case indicates the position of cleavage).

Microbial origin	Enzyme	Recognition site (two fold symmerty)
Escherichia coli	*Eco*RI	↓ 5′ GAA TTC 3′ 3′ CTT AAG 5′ ↑
Haemophilus influenzae	*Hind*II	↓ 5′ GTPy PuAC 3′ 3′ CAPu PyTG 5′ ↑
	HindIII	↓ 5′ AAG CTT 3′ 3′ TTC GAA 5′ ↑
Haemophilus parainfluenzae	*Hpa*I	↓ 5′ GTT AAC 3′ 3′ CAA TTG 5′ ↑
	HpaII	↓ 5′ CC CG 3′ 3′ GG CC 5′ ↑
Haemophilus aegypitus	*Hae*III	↓ 5′ GG CC 3′ 3′ CC GG 5′ ↑
Moraxella spp.	*Msp*I	↓ 5′ CC GG 3′ 3′ GG CC 5′ ↑
Proteus vulgaris	*Pvu*I	↓ 5′ CGA TCC 3′ 3′ GCT AGC 5′ ↑
Bacill amyloliquefaciens	*Bam*HI	↓ 5′ GGA TCC 3′ 3′ CCT AGG 5′ ↑
Streptomyces alb	*Sal*I	↓ 5′ GTC GAC 3′ 3′ CAG CTG 5′ ↑
Providencia stuortii	*Pst*I	↓ 5′ CTG CAG 3′ 3′ GAC GTC 5′ ↑

Technique of Restriction Mapping

Genetic maps are prepared using recombination frequencies. However, the recombination frequencies, which are the function of distances between genes, are not completely independent of (i) the nature of mutants used, (ii) the position of these mutants on chromosomes, (iii) the genetic background, (iv) the environmental conditions, and (v) a variety of other factors. Consequently, the distances between genes on a genetic map may not correspond to the distances between them on the DNA molecule of which they are a part at the molecular level. Gaps may also be present on a genetic map due to non-availability of mutants in that region.

At the molecular level, the fine structure of a gene can be studied through determination of nucleotide sequence of the concerned DNA segment. Sometimes this is done through isolation of a DNA segment corresponding to a gene followed by sequencing of the DNA segment, which needs time and energy. Instead we can prepare a map of the DNA by its cleavage at specific sites with the help of restriction endonucleases, which recognize very short specific DNA sequences and cleave the DNA at these specific sites. These sites of cleavage can be identified and mapped to give rise to a restriction map.

A restriction map thus consists of a linear sequence of sites, each for a specific enzyme and the distances between them are measured in terms of number of base pairs of DNA. This technique can be used both in prokaryotes and eukaryotes, although all regions of chromosomes cannot be easily mapped in eukaryotes. Different steps involved in restriction mapping include the following.

Restriction Cleavage and Gel Electrophoresis : If we take a particular DNA molecule or DNA sample, digest it with a specific restriction enzyme and then subject the sample to gel electrophoresis (a technique in which DNA digest is loaded on

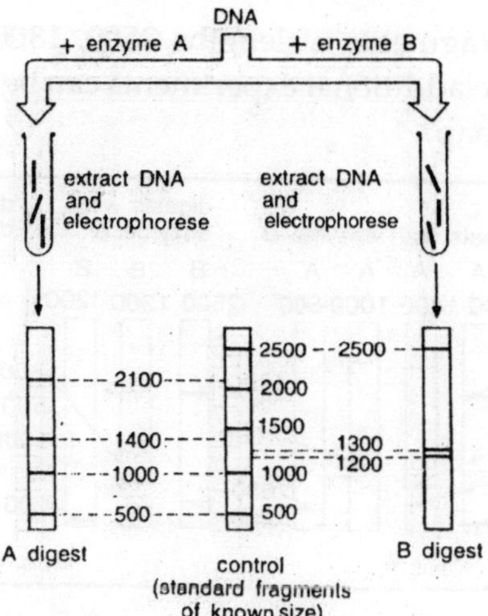

Cleavage of DNA by two restriction endonucleases A and B into fragments which can be separated by agarose gel electrophoresis.

gel slab and the DNA fragments move under the influence of an electric current), we will notice a series of bands on the gel slab or cylinder. The bands can be observed under ultraviolet light after staining with ethidium bromide. The position of different bands will depend on DNA fragment size, such that smaller the fragment, more rapidly will it move, and longer the fragment, more slowly will it move. It will mean that the fragment away from the loading site will be smaller and those close to the loading site will represent longer DNA fragments. The gel can be calibrated by using a mixture of DNA fragments of known lengths, so that the positions of the bands of known lengths on this standard gel can be compared with the positions of bands in the experimental DNA digest. The fragment length for each band of DNA digest can thus be determined sometimes using a computer device. In Figure, we have shown the results of digestion of 5000 bp long hypothetical DNA molecule digested separately by two enzymes A and B. As shown in the figure, the enzyme A cleaves the DNA into four fragments of lengths 2100, 1400, 1000 and 500bp, while the enzyme B cleaves it into three fragments of lengths 2500, 1300, 1200bp. These data with some additional experiments can be used to generate a restriction map.

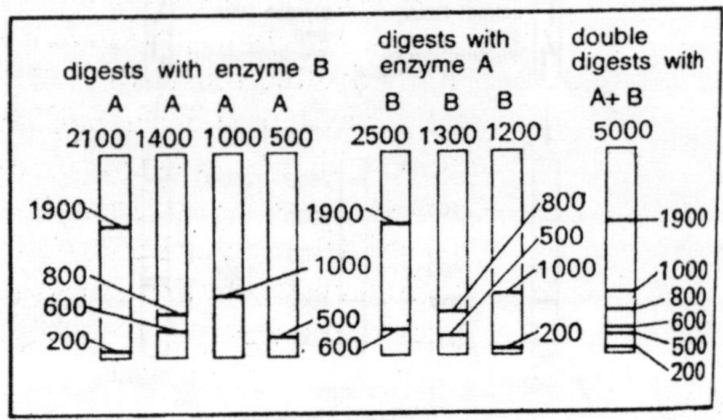

Technique of double digests to determine cleavage positions of DNA due to one enzyme with respect to those due to another enzyme (the two enzymes are A and B); the four gels on the extreme left show the results of electrophoresis after digesting with enzyme B, each of the four fragments obtained after digestion with A; three central gels represent results of digesting with enzyme A, each of the three fragments earlier obtained with enzyme B and the solitary gel on the right shows the results of digesting the intact DNA with both the enzymes simultaneously.

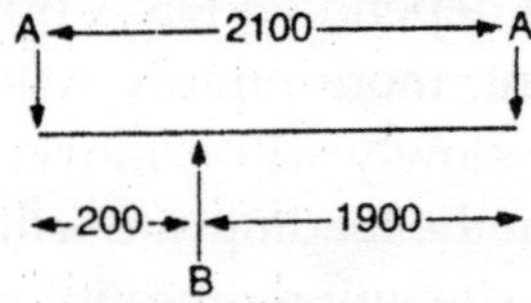

The sites of cleavage by enzymes A and B in one of the four fragments, 2100 bp long, obtained due to cleavage by enzyme A; the fragment has a site for enzyme B at 200 bp length from one end.

Construction of a Restriction Map

The data on DNA digestion by more than one endonucleases as discussed above can be utilized to arrange the sites of cleavage in a defined order. This can be done by several methods and one of them is illustrated in the figure shown. This method involves successive digests with two individual enzymes, where we extract each fragment produced in the individual digests with either enzyme A or enzyme B and then cleave it with the other enzyme. The original DNA sample is also digested by a mixture of both the enzymes to confirm the results of individual successive digests. In Figure, we have shown results of such an exercise. When the fragment A-2100 is obtained and digested with enzyme B, it is cut into two fragments of 1900 bp and 200 bp. Fragment B-2500 (obtained from individual digest with enzyme B) is similarly oil by enzyme A into two

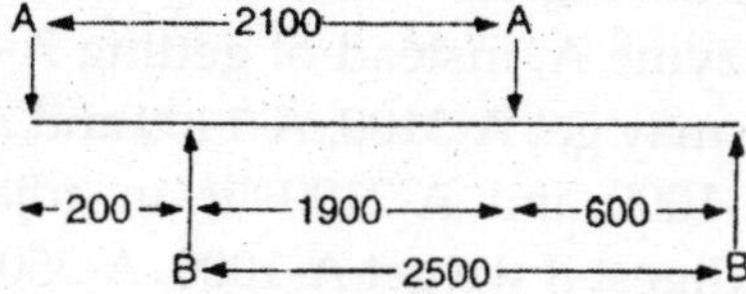

Reconstruction of restriction map showing restriction sites of enzymes A and B in two overlapping fragments (2100 bp fragment obtained due to enzyme A and 2500 bp fragment obtained due to enzyme B).

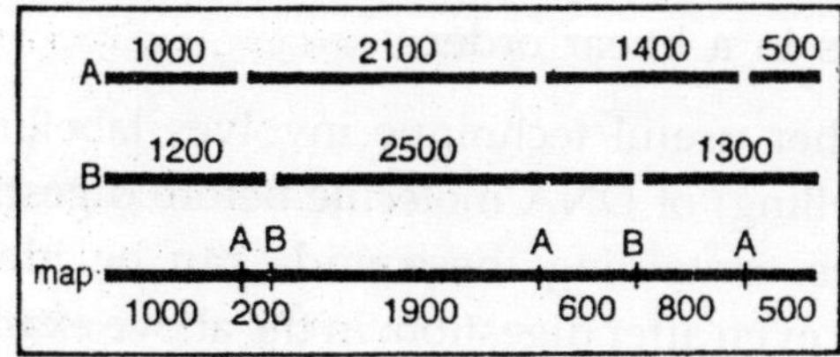

A restriction map, in the form of linear arrangement of cleavage sites; the upper two figures show sites due to individual enzymes and the bottom figure shows arrangement of cleavage sites of both enzymes (see text for details); sites of more enzymes can be added to this map by same technique.

fragments of 1900 bp and 600 bp, suggesting overlap of A-2100 and B-2500 (Fig.), in the region of fragment of 1900 bp which is obtained by one cut with enzyme A and the other by enzyme B. The DNA fragment of 1900 bp is also available, when DNA is digested with a mixture of both enzymes A and B (double digest). It may be seen in figure that in reciprocal digests (A followed by B and B followed by A) and double digests, same fragments are available. Using this information, we can find the overlapping regions in A and B digests and find out the sites of cleavage by A and B. This will then allow us to prepare the restriction map as shown in Figure.

Employing Partial Digests, End Labelling and Hybridization in Restriction Mapping : The above technique of individual and double digests can be supplemented with other techniques for actual construction of maps. For instance, by permitting incomplete digestion, fragments longer than those obtained by complete digestion may be obtained. These will be called partial digests. The example used earlier in partial digests with enzyme A, instead of getting A-1000, A-2100, A-1400, A-500, we may get A-3100, A-1400 and A-500, which will suggest that A-1000 and A-2100 lie in adjacent regions. In another partial digest if we get A-1000, A-3500 and A-500, this will suggest that A-2100 and A-1400 are adjoining, which will mean that A-1000 and A-1400 are found on the two sides of A-2100. This technique will thus allow the arrangement of fragments in a linear order.

Another useful technique involves labelling of the ends (end labelling) of DNA molecule before digestion, so that the fragments containing these ends can be identified due to labelling, even after digestion. In the above example, using end labelling, if A-1000 and A-500 are found to be radioactively labelled, these will be present at the two ends of the restriction map. Some of the features of a restriction map can also be confirmed by nucleic acid hybridization. For instance, in the above example, if A-2100 and B-2500 have an overlapping region of 1900bp, they should hybridize with each other, which will confirm overlapping.

The above technique will help in accurate completion of a restriction map, but will require that we have a complete set of restriction fragments which make the entire DNA region being mapped. Once the restriction map (which is a physical map) is ready, it can be compared with the genetic map. Although large changes, already located on the genetic map, can be easily located on the restriction map, but point mutations cannot always be easily located on the restriction map, since the restriction sites often do not change due to mutation. In such a situation, one may like to determine the nucleotide sequences of individual fragments and compare them in normal and mutant individuals. Since sequencing of nucleotides is more laborious and cannot be easily undertaken for the whole genome, another technique of using a variety of molecular genetic markers (e.g., RFLPs, RAPDs, STs, STMS, AFLP, etc.) has been utilized during the 1980's and 1990's to generate genetic linkage maps in human, mouse, rat, pig, sheep, fruitfly *(Drosophila)*, nematode *(Co-enorhabditis elegans)*, yeast, maize, wheat, rice, barley, tomato, potato, lettuce, rice, pepper, pea, etc.

Building Process of Chimeric DNA

Chimeric DNA molecules are produced by inserting a foreign DNA segment into the DNA molecule of vector. These chimeric DNA molecules are the means of cloning genes or other small DNA sequences. The construction of chimeric DNA, as we know now, involves joining of the broken ends of the vector with the two ends of the sequence to be cloned. There are several methods available for the construction of this chimeric DNA.

Plasmid Vectors' Cloning Process

Cloning DNA fragments with homologous cohesive ends. This is the most common method, where due to the presence of a palindromic sequence, a restriction enzyme causes staggered cuts producing short complementary sequences at the two ends of (a palindromic sequence is one, which has complementary sequences at the two ends of a single-strand, e.g. ATC GAT). This can be illustrated using the example of enzyme *EcoRl*, which cuts the sequences at specific positions (Fig.). When another DNA molecule is similarly cut by the same enzyme, similar sticky ends having same sequences in the single stranded ends will be produced, so that when it is mixed with the previous molecule similarly treated, the two will anneal producing a chimeric DNA or chimeric plasmid, if plasmid DNA is involved (Fig.). Enzyme DNA ligase helps in joining the bonds at the cut ends of the two molecules. This method of cloning is commonly employed with pBR322 and pUC vectors.

The above technique has the advantage of regenerating two restriction sites *(e.g. EcoRl* sites) in the chimeric DNA, so that the foreign DNA segment can be retrieved rather easily from the cloned copies of chimeric DNA by cleavage again with the same enzyme. There are also following disadvantages with this technique, (i) The cleaved ends of two vector molecules or those of two foreign DNA segments may join end to end before the foreign DNA gets inserted. Therefore, while isolating chimeric DNA, one will have to select chimeric molecules each having only a single insert. This can be achieved by separating molecules of different sizes by gel electrophoresis. This difficulty can be overcome, *firstly,* by adjusting the concentrations of two types of DNA and *secondly,* by removing 5' phosphate groups in the linearized vector, (ii) The recognition site, particularly in the sequence to be cloned may not lie at a convenient position, so that sometimes only a part of the desired segment will be inserted.

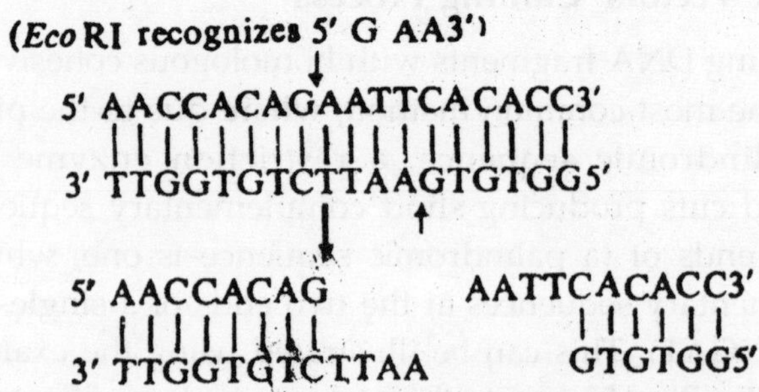

Staggered cleavage of a double-stranded DNA fragment, producing' single-stranded sticky ends

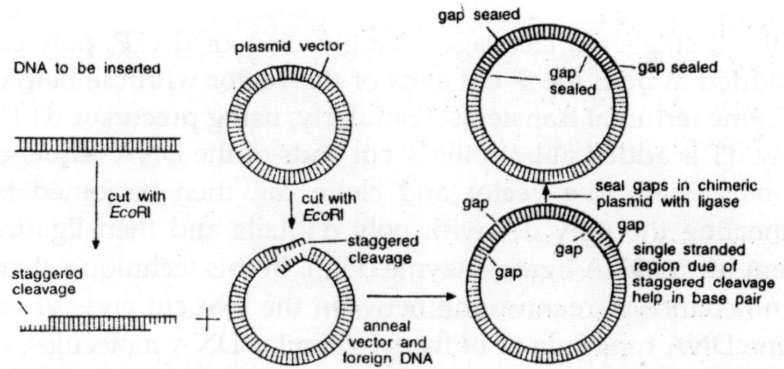

Cloning of a DNA fragment in a plasmid vector (e.g. pBR322) by staggered cleavages, both in vector DNA and foreign DNA to be cloned.

Directional Cloning of DNA Fragments with Heterologus Protruding Termini. This is the most efficient method for cloning, and any other method should be tried only when this method can not be used. In this method, the fragments to be cloned are generated by digestion with two different restriction enzymes. Similar heterologous termini are generated in the plasmid vector by removing a small fragment due to cleavage with two different enzymes. In this method, since the two termini are not compatible, self ligation of either the linearized vector or the foreign DNA fragment leading to circularization is eliminated. Moreover, chimeric DNA molecules will contain each only a single copy of the insert DNA oriented in a defined direction. Recombinant molecules with two or three copies of insert will be very rare (this is unlike the case with cloning of DNA fragments with homologous cohesive ends).

As an illustration of directional cloning, pUC19 can be digested with *BamHl* and HindIII and the resulting large fragment can be purified, removing the small fragment. The vector can then be ligated (using T4 ligase) to a segment of foreign DNA that contains termini compatible with those generated in the vector due to *BamHl* and *Hindlll.* The steps involved in this directional cloning are shown in Figure.

Adding Poly dA at 3′ Ends of the Vector and Poly dT at the 3′ Ends of DNA Clone : In this method, DNA can be cut at the desired position both in the vector and in the clone without staggered cleavage. Using precursor dATP, poly dA is added at both the 3' cut ends of the vector with the help of enzyme terminal transferase. Similarly, using precursor dTTP, poly dT is added at both the 3' cut ends of the DNA sequence to be cloned. The vector and clone can then be joined by annealing the poly dA with poly dT tails and then ligating them using DNA ligase enzyme (Fig.). In this technique, there is no chance of reannealing between the two cut ends of the same DNA molecule or of the two similar DNA molecules, so that one of the disadvantages of the first method has been overcome in this case. However, in this technique it will not be easy to retrieve the cloned DNA, because the recognition site of the enzyme has been lost in this case due to insertion of poly dA and poly dT. But if poly dG and poly dC are used instead (which serve the same purpose), cloned sequence can be retrieved, since, poly dG and poly dC regenerate the recognition site for the enzyme *Pstl.*

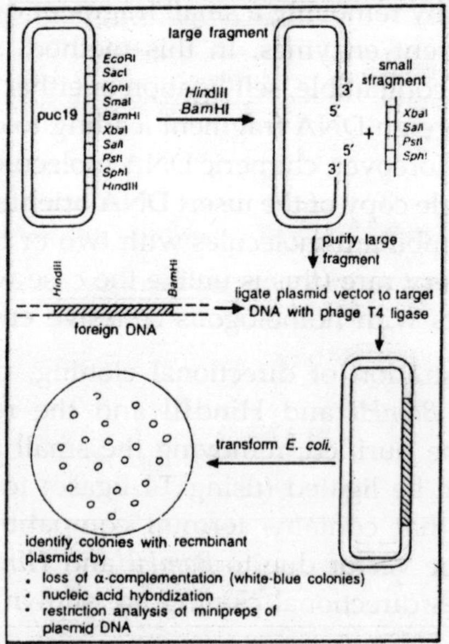

Cloning of a DNA fragment in a plasmid vector by cleavage of both vector DNA and the foreign DNA by two different enzymes, producing non-compatible termini.

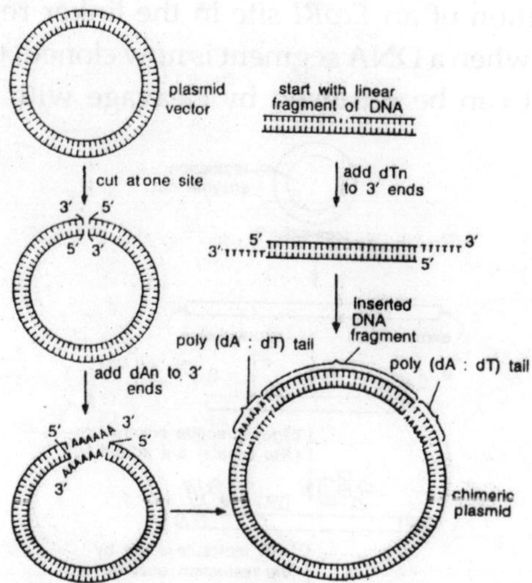

Cloning of DNA segment in a plasmid vector by poly dA : poly dT tailing technique.

Blunt End Ligation by T4 DNA Ligase : In this technique, a restriction enzyme is used to cut a duplex DNA at the same place in both the DNA strands. The broken ends are then used for joining with the two ends of another DNA molecule irrespective of the sequences present at the broken ends of the two DNA molecules. The T4 DNA ligase is used for this joining reaction. The disadvantage of this technique is that any two broken ends may join including those belonging to the same DNA molecule. This leads to the production of a variety of products and one will have to select the desired product from a mixture of products.

The blunt end ligation is used for developing a method in which the DNA to be cloned can be easily retrieved whenever *(BC-45)* required. This method makes use of short DNA duplexes (linkers), that contain *EcoRI* palindrome or some equivalent palindrome, which being small in size, can be synthesized chemically. These linkers can be linked to the blunt ends of vector DNA by blunt end ligation (Fig.). This will allow the creation of an *EcoRI* site in the linker region of the vector, so that when a DNA segment is now cloned, the inserted DNA segment can be retrieved by cleavage with EcoRI.

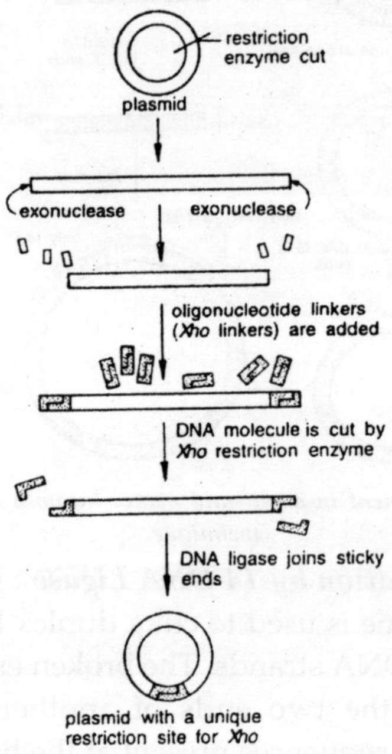

Addition of a "linker" (carrying a restriction site) to a vector molecule.

Phage λ s *and Cosmid Vectors Cloning*

Cloning in Phage λ Vectors : Phage λ vectors are used for a variety of purposes including the following : (i) subcloning of DNA fragments that are propagated in vectors with large capacity (e.g. cosmids) and (ii) construction of genomic and cDNA libraries.

Cloning in phage λ vectors involves the following steps : (i) digestion of phage λ with a restriction enzyme (digestion with two enzymes, one followed by the other is sometimes carried out to remove the stuffer; this is necessary in case of only some of the replacement vectors) and purification of phage λ ; purified arms of several popular λ vectors including those for EMBL3 and EMBL4, and those for λ *gt10* and λ gtll are also available commercially, so that if the arms are procured commercially, this step involving digestion and purification can be dispensed with; (ii) ligation of λ arms to fragments of foreign DNA to be cloned; (iii) packaging of recombinant phage λ *in vitro*, utilizing the packaging extract that is available commercially or can be prepared in the laboratory (packaging extract is a lysate of phage λ infected *E. coli* cells and contains empty phage heads, unattached phage tails and phage-encoded proteins needed for DNA packing); this is followed by infection of *E. coli* competent cells by packaged recombinant λ vectors, although rarely transfection with naked λ DNA without packaging can also be tried; (iv) multiplication of transformed *E. coli* cells and (v) lysis of *E. coli* cells to release recombinant phage, forming plaques and recovery of cloned DNA segment.

Cosmid Vectors Cloning : Cosmids were developed to allow cloning of large DNA fragments (~40 kb) in plasmid/ phage λ like vectors. The vector is cut with a restriction enzyme and mixed with pieces of DNA to be cloned (Fig.). The cleaved vector and the foreign DNA fragments are made to join to produce concatameric molecules with the help of DNA ligase (concatamers = several molecules joined together end to end). The ligation mixture is then mixed with a packaging extract (commercially available) to allow packaging of chimeric cosmid DNA into phage heads.

The cosmid containing phages are infectious. After infection, the cosmid circularizes, and multiplies in *E. coli* cells, just like a plasmid. Like a plasmid, a cosmid is also small in size (5-8 kb), but like a phage it can be packaged in λ head, thus permitting cloning of large DNA fragments (~40 kb; λ phage genome = 50 kb and therefore can accommodate no more than ~20 kb).

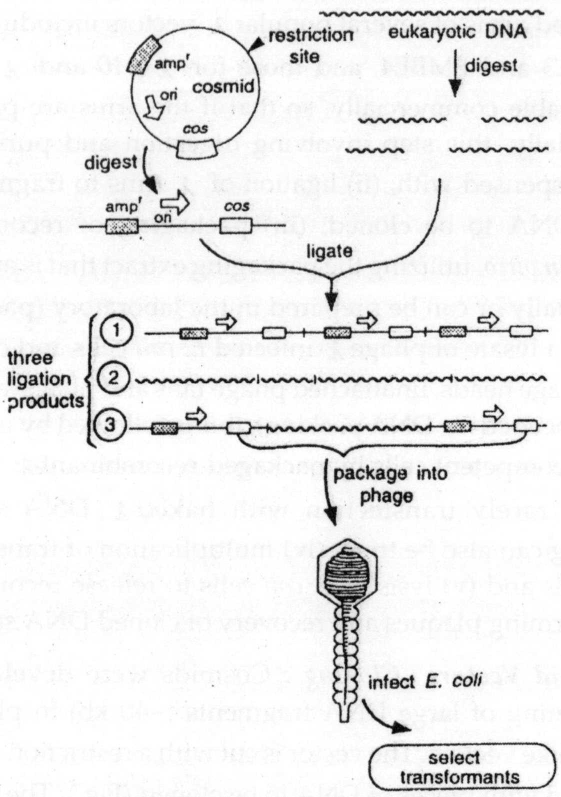

Cloning of a large DNA fragment (-40kb) in a cosmid vector.

Cloning Vectors' Hosts

Bacteria as Hosts : Bacteria are often used as hosts for multiplication of chimeric vectors or recombinant DNA molecules. Therefore, a bacterial strain should be available, which should be modified or selected for high transformation efficiency to be used as a host.

Bacteria can be classified into, (i) gram negative bacteria e.g. *Escherichia coli. Pseudomonas, Rhizobium* and (ii) gram positive bacteria e.g. *Bacillus subtilis* and *Streptomyces* sp. Ordinarily same plasmid can not maintain itself in stable form in both classes of bacteria and, therefore, plasmid vectors for two classes usually differ. While gram negative bacteria can be used as hosts for plasmids, phages and cosmid vectors, gram positive bacteria can be used as hosts for only plasmids. Chimeric vector is inserted into bacterial cells and transformed bacterial colonies are selected and used for multiplication in suspension cultures. The chimeric DNA is retrieved, whenever required.

Eukaryotes as Hosts : Since chromosomes found in the nucleus of eukaryotes are separated from the rest of the cell through nuclear membrane, and since many of the genes are split genes with exons and introns, genetic engineering with eukaryotes required new methods and tools. When eukaryotic genes are cloned in prokaryotes, the split genes can not be correctly expressed, because prokaryotes do not have the equipment necessary for splicing out the RNA transcribed from the introns of a gene. In view of this, eukaryotic cells may sometimes be needed for cloning and particularly for expression of cloned eukaryotic genes.

Among eukaryotes, DNA cloning has been done in cultured cells of yeast, mouse, human and to some extent even in those of some higher plant species. In yeast, a plasmid called 2 μ DNA (63bp) is found, which is an appropriate cloning vehicle. Yeast artificial chromosome (YAC) vectors can also be used for cloning in yeast cells.

An efficient transformation method is also available, which involves protoplast production followed by PEG-directed introduction of DNA into plant protoplasts. In animal cells like mouse cells, special animal viruses were used as cloning vehicles. Simian virus 40 (SV40) is one such virus, in which globin gene could be integrated. This gene integrated in SV40 could be transcribed and translated in mouse kidney cells. A variety of human cell lines in culture is also used for cloning of foreign DNA, using some animal viruses as vectors.

Molecular Probes

Molecular probes are small DNA or RNA segments that recognize complementary sequences in DNA or RNA molecules and thus allow identification and isolation of these specific DNA sequences from an organism.

Antibodies are also occasionally used as probes to recognize specific protein sequences. Although, initially, these probes were developed and used for genetic engineering research but are now frequently used for a variety of purposes including diagnosis of infectious diseases, identification of food contaminants, variety of microbiological tests, forensic tests (e.g. fingerprinting of murderers or rapists, etc. 'Probes can also be used to identify strains' of an organism, e.g. varieties of a crop species (a plant breeder likes to have a quick test to identify his variety to maintain his patent or breeder's right, so that a competitor may not use it in another name leading to infringement of rights).

For basic studies in molecular biology laboratories, the molecular probes are frequently used for identification and isolation of genes or related sequences. In theory, any nucleic acid (or rarely protein, e.g. antibody) can be used as a probe, provided it can be labelled to permit identification and quantitation of the hybrid molecules formed between the probe and the sequence to be identified. In practice, double and single-stranded DNAs, mRNA and other RNAs synthesized *in vitro* are all used as probes.

DNA/RNA probe assays are faster and sensitive, so that many conventional diagnostic tests for viruses and bacteria involving culturing of the organism, are being fast replaced by antibody and DNA probe assays. While culture tests can take days or even months, molecular probe assays can be performed within few hours or minutes. Therefore, the use of DNA probes has become today's most sophisticated and sensitive technology for a variety of uses involving biological systems both in basic and applied studies including their commercial use.

Preparing Probes

Genomic DNA Probes : Genomic DNA probe can be either specific for identification of specific homologous sequences in the genome (for diagnostic kits and isolation of genes, etc.) or alternatively these may be random anonymous probes of unknown sequences, which are used for the study of DNA polymorphism and construction of molecular maps without caring about the genomic DNA sequences to which this polymorphism belongs. In other words, genomic DNA probes can be specific or random. Same will also be true for cDNA probes discussed in the next para. Usually these probes may consist of DNA fragments which can be as small as 50-200 base pairs to as long as several kilobase pairs long. Although specific probes may sometimes also belong to repetitive DNA, like ribosomal DNA, or minisatellites/microsatellites, majority of probes, used for the study of genomic DNA polymorphism and construction of molecular maps, belong to unique DNA. For the preparation of random genomic DNA probes, genomic DNA is first digested with a restriction enzyme, that is a frequent cutter (enzymes with a recognition site that is only four base pairs long are frequent cutters because four bases long sites will be relatively more frequent than those with six or eight bases).

The digested DNA may be subjected either to gel/ polyacrylamide electrophoresis or to sucrose density gradient centrifugation. In either case, fragments in a specific range (200-500 base pains) are separated, cloned and characterized to be used as molecular probes. If genomic DNA probe of a specific DNA sequence (e.g. ribosomal DNA, or specific gene) is needed, then such a fragment will have to be detected in the DNA digest or in a genomic library by any of a variety of methods that are available. Specific genes can also be used as molecular probes, particularly for the purpose of diagnosis.

A DNA fragment to be used as a probe is multiplied in bacterial cells with the help of a vector. From transformed bacteria, the chimeric vector can be obtained and used in one of the following ways : (i) It may be used directly as probe (the presence of vector DNA will not interfere in probe assays, so that it is not really necessary to remove it), (ii) The cloned segment may be separated (or retrieved), by using the same enzyme which was used for cloning. In the latter case, cleaved chimeric vector DNA will be again electrophoresed for separating the inserted segment on the gel. This inserted segment thus retrieved can now be used as a probe, (iii) The chimeric DNA may also be used for PCR, using flanking sequences as primers; the PCR product can be separated and used as a probe.

cDNA Probes : A DNA sequence corresponding to a part of a specific gene(s) (an oligonucleotide) can be obtained by reverse transcription of mRNA. The cDNA thus obtained can be cloned and used as a probe, in a manner similar to that discussed above for genomic DNA probes.

The Role of Synthetic Oligonucleotides as Probes : DNA probes with known nucleotide sequences can also be synthesized chemically using automated DNA synthesizers now available commercially. These synthetic probes will be efficient only when they are not more than 20 to 40 nucleotides in length. Use of SSRs or simple sequence repeats (like di-, tri- or tetranucleotide repeats) for oligonucleotide fingerprinting is one example of the use of synthetic oligonucleotide probes.

RNA Probes or Riboprobes : High specific activity RNA probes or riboprobes may also be synthesized from DNA templates cloned in expression vectors. This is achieved through RNA synthesized *in vitro* (on DNA inserted in vector) and labelled simultaneously employing specific RNA polymerase and rNTPs, which are supplied by commercial firms in a kit for riboprobe preparation. RNA probes offer several advantages over DNA probes, since these are single stranded, and provide improved signal or hybridization blots. However, widespread presence of ribonucleases creates some problems in their preparation and use, because these ribonucleases will cause degradation of RNA which is synthesized. Therefore, extreme care is needed in the preparation of riboprobe, keeping all glassware ribonuclease free (no glassware should be touched with bare hands, because ribonucleases may be available on our skin).

Probes' Labelling Process

The detection of homologous sequences after hybridization with the probe is like finding a needle in the haystack. Therefore, for the success of DNA probe assay, it is necessary to develop simple, safe and sensitive techniques for their use. As probes transmit no signal of their own, they have to be (i) either labelled with radioactive isotopes or (ii) methods are used to couple non-radioactive signal molecules to the probes without impairing the hybridization ability of these probes. These signal molecules may include fluorescent antibodies and enzymes that produce colour changes in dyes and chemiluminescent catalysts (Fig.).

Probes of Radio-Labelled Type : Traditionally radio-actively labelled probes are used for a variety of experiments. ^{32}P is the most commonly used radioisotope, although other labels such as ^{3}H, ^{35}S and 1251 have also been used. Conventional labelling replaces a proportion of nucleotides in a nucleic acid molecule with ^{32}P derivatives (e.g. $^{32}PdCTP$) or adds ^{32}P to the end of the molecule. At least three methods are available for preparation of a labelled probe from a double-stranded DNA: (i) nick translation involves creation of nicks in the probe DNA followed

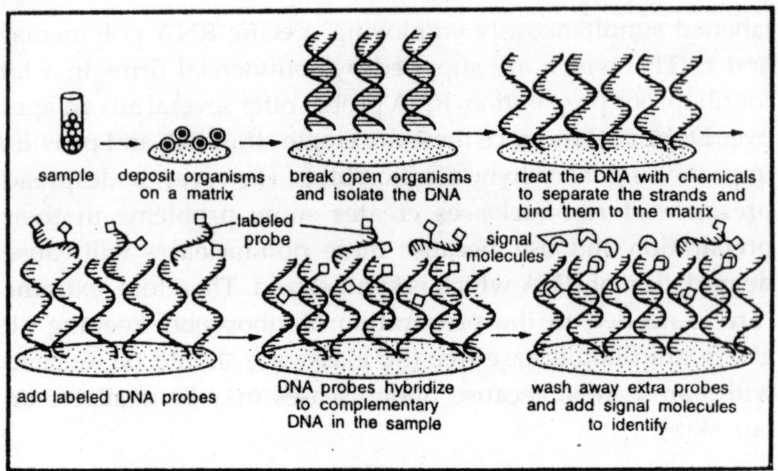

Use of labelled DNA probe for identification of homologous DNA sequence through DNA hybridization

by extension of the broken ends using a labelled deoxyribonucleotide with the help of DNA polymerase I (or 'Klenow fragment' of this enzyme); (ii) oligonucleotide labelling involves the use of short random oligonucleotides, which are used as primers for copying the probe DNA in presence of labelled deoxyribonucleotides; (iii) riboprobe preparation involves synthesis of labelled RNA with the help of RNA polymerase, using DNA probe as template, in presence of a labelled ribonucleotide.

Single-stranded DNA or RNA probes, which consist of only one of the two complementary strands of a given sequence, are also used and have an advantage over the more traditional double stranded probes. The absence of the complementary strand in a single-stranded probe eliminates the possibility of forming non-productive hybrids composed of reannealed probe. These probes are particularly useful in experiments involving hybridization between RNA and DNA, as required in nuclease S1 mapping (S1 enzyme will degrade unhybridised single-stranded DNA and will allow mapping of exon sequences that are found in mRNA).

Riboprobes described above are single-stranded probes, but single-stranded DNA probes can also be synthesized. A DNA segment cloned in a phagemid or M13 vector is used as template for the synthesis of radiolabelled single-stranded DNA. A sequence complementary to a sequence of the vector is used as a primer and the four dNTPs, of which one should be labelled, are used for DNA synthesis by Klenow fragment of *E. coli* DNA polymerase I.

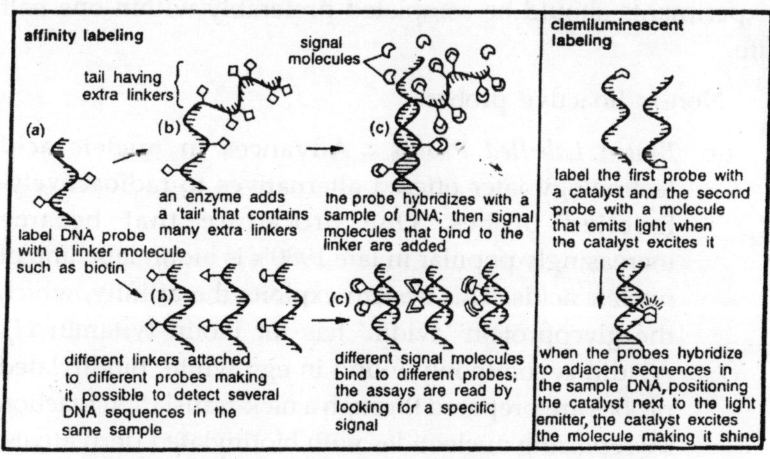

Non-radioactive labelling of probes, and the use of three-labelled probes in DNA Hybridization.

In all above cases, labelling involves incorporation of labelled dNTPs at internal positions of DNA. However, in some cases, when the probe (double stranded or single stranded) is small (~20 mer), end labelling of 5' or 3' termini of DNA is needed. Such end- lebelled probes are used (i) for nuclease-Sl mapping of RNA, (ii) as radio-labelled primers in primer extension reactions and (iii) for in-gel hybridization (e.g. in oligonucleotide fingerprinting. The double-stranded DNA is generally labelled at the 3' termini of both the strands using either *E. coli* Klenow fragment or T4 DNA polymerase, and the oligonucleotides (single stranded) are labelled at their 5' termini using T4 polynucleotide kinase.

After hybridization, with labelled probe, hybrids are detected by autoradiography (Fig.). ^{32}P has the advantage over other radioisotopes, since it has high specific activity. However, in general, radioisotopes have some disadvantages. They are difficult to handle and expensive to dispose off. Detection by autoradiography, while sensitive, may take a long time if there are few counts in the hybrid. Furthermore, radioisotopes have a short half life (e.g. ^{32}P has a half-life of 14.3 days) and therefore experiments should be completed preferably within one half life.

Non-radioactive probes:

(a) *Biotin Labelled Probes* : Advances in nucleic acid technology later offered alternatives to radioactively-labelled probes. One procedure that became increasingly popular in late 1980's is biotin labelling of nucleic acids. This system exploits the affinity, which the glycoprotein 'avidin' has for biotin (vitamin H). Avidin is commonly found in egg white. Biotinylated probes are prepared through a nick-translation reaction by replacing nucleotides with biotinylated derivatives. After hybridization and washing, detection of hybrids is done by a series of cytochemical reactions which finally give a blue colour whose intensity is proportional to the amount of biotin in the hybrid. There are several advantages of using biotinylated probes. For example, these assays employ non-toxic materials, whose half-life is longer. These probes can be prepared in advance in bulk and stored at -20°C for repeated uses. Detection of hybrids is much faster than in case of radioactive probes. A limitation of this technology is that a very small probe (20 nucleotides) contains only a small number of biotinylated sites, limiting the intensity of signal obtained. It has been solved by adding long 'tails' of biotinylated nucleotides to the probes through enzymatic methods. Sometimes the probe does not need to be labelled with biotin, but only coupled with a tail. Another disadvantage of biotin-labelled probes is that the cytochemical visualization reactions lead to precipitation of insoluble material which cannot be removed and therefore, the filter cannot be reused, while in radiolabelled probes, the filters can be repeatedly used for hybridization with a number of probes, one at a time.

(b) *Probes Labelled with Digoxigenin* : Digoxigenin is another chemical derived from plants and used for non-radioactive labelling of probes. An antibody associated with an enzyme (anti-digoxigenin-alkaline-phosphatase conjugate) is used for the detection of the presence of digoxigenin. The probe may be labelled with digoxigenin-11-dUTP supplied with a digoxigenin kit (these kits are available from any commercial firm, e.g. Boehringer Mannheim). The labelled and denatured probe may be used for hybridization with denatured DNA on Southern blots or for in-situ hybridization. After hybridization the membrane or the slide may be washed and transferred into detection buffer containing 20 Hg/ml of anti-digoxigenin fluorescein and 5% (w/v) BSA (bovine serum albumin). The system is incubated for one hour at 37°C, and then the membrane/slide is washed in detection buffer three times (8 min each at 37°C). Alkaline phosphatase activity can be detected using 0.17 mg/ml BCIP (5-bromo-4chloro-3 indolyl phosphate) and 0.33 mg/ml NBL (nitroblue tetrazolium) as dye substrate.

(c) *Substitutes to Biotin and Digoxigenin Labelling*. The techniques of non-radioisotopic labelling have been further expanded and new methods have been devised for attaching other ligands (e.g. hapten determinants, 2, 4-dinitrophenol, arsenate derivatives, etc.) to nucleotides without hampering their ability to be incorporated into DNA (see Figure for some details). These alternatives required binding of attached ligands to specific proteins that can be tagged with enzymes or fluorescent molecules. It is possible to monitor many probes simultaneously by using several different ligands since each ligand would yield a different signal. A chemiluminescent (light emitting) probe system has also been developed in which two different probes complementary to a continuous segment of DNA hybridize to adjacent segments of a gene. The first label is a chemiluminescent complex that emits light at a specific wavelength; this emission excites the label molecule on the second probe to emit light at a different wave length which can be detected using a photo-multiplier device (Fig.). This process called non-radioactive energy transfer can occur only if the two probes hybridize correctly and the two labels are close to each other. This system has great fidelity and provides basic technology for a homogeneous assay. In addition, DNA does not need to be immobilized and no washing steps are necessary which may be an additional advantage for large-scale testing.

DNA Probe Signals' Amplification : During 1980's and early 1990's, DNA probe assays and immunoassays (using antibodies) competed with each other for a variety of purposes. While DNA probes have an advantage of detecting the presence of gene rather than its product, the immunoassay has the advantage of getting amplified target (the gene product) firstly due to transcription (several mRNA molecules can be synthesized on single copy of gene) and secondly due to translation (several protein molecules can be produced from same mRNA molecule). The amplified target gives an amplified signal. Therefore, if DNA probes have to compete with monoclonal antibodies involving immunoassay, DNA probe signals need to be amplified. Following devices have been used for signal amplification, (i) Probes have been developed for rRNA, since rRNA genes are present in thousands thus amplifying the target. (ii) Polymerase chain reaction (PCR) technology (which utilizes a thermostable polymerase) is used to generate millions of copies of target DNA sequence, which can then be identified by standard methods, (iii) Signal-generating capacity of the probe can also be increased by any one of the following methods :

(i) concentrating more label at the site of target molecule by attaching multiple enzyme molecules to each of its DNA probe or

(ii) by using multiple probes, or

(iii) by attaching multiple enzymes to each of the multiple probes,

(iv) Ampliprobe technology uses multiple secondary probes (each having one or more enzyme molecules) that hybridize to multiple target specific primary probes.

The secondary probe is independent of target but helps in amplifying the signal. These approaches have been described as 'Christmas Tree' or 'Christmas Forest' approaches.

The above amplification systems have played a significant role in the use of DNA probes for a variety of purposes.

Methods Employed in Molecular Probing

Separation of DNA Fragments using Agarose or Polyacrylamide Gel Electrophoresis (PAGE) : When genomic DNA, extracted from any tissue of a plant or animal species, is digested with a restriction enzyme, it is cleaved into segments. The segments of different sizes can be separated through gel electrophoresis before a molecular probe is used to detect the segments which have sequences similar to those in the probe. Gel electrophoresis involves movement of fragments or molecules under a high voltage electric current. The mixture of DNA fragments is loaded in a well-created on one edge of the gel. The gel may be a cylinder or a slab (usually a slab for cloning experiment.), about 10 cm long and 0.5 cm thick. The rate of movement of fragments is inversely correlated with the size of fragments or molecules, so that heavier fragments will remain closer to the site of loading and the lighter fragments will move away. Fragments of different sizes will appear as bands on the gel and can be examined or isolated for further study. More often agarose gels are used, but for separation of fragments differing by few base pairs, polyacrylamide gels are used. Polyacrylamide gels are more commonly used for DNA-sequencing experiments.

Separation of large DNA molecules (whole chromosomes) using PFGE. The technique of agarose/polyacrylamide gel electrophoresis is used for separation of DNA molecules of different sizes. However, DNA molecules of large size could not be handled in this technique. A new technique called pulsed

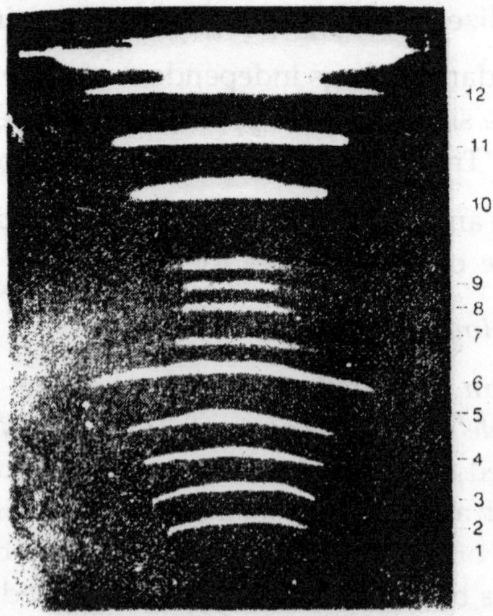

A gel with 12 bands representing 15 chromosomes of yeast as resolved by PFGE (the three doublet bands i.e. 5, 10 and 11 can be resolved on other gels).

field gel electrophoresis (PFGE) has been used, for separation of large-sized DNA molecules, sometimes representing whole chromosomes. Using this technique, separation of DNA molecules belonging to each of the 16 individual chromosomes of yeast has become possible (Fig.). In this technique, short pulses of electricity are used in two different directions and DNA is embedded and used in the form of agarose plugs to avoid fragmentation of large DNA molecules. Using the technique of PFGE and a more refined technique CHEFE (countour clamped homogeneous electric field electro phoresis), genomes of several fungi could be resolved into chromosomal bands and used for mapping of DNA sequences on specific chromosomes.

Blotting of Southern, Northern and Western : A mixture of DNA, RNA or protein fragments can be separated by gel electrophoresis and the separated bands can be stained and visualized directly in the gel. However, to confirm the identity of these bands or to find similarity of one or more of these bands with a known and available molecular probe, it is possible to hybridize these bands with a labelled probe. To facilitate this hybridization, the bands are often transferred to a nitrocellulose membrane through a technique described as blotting. When DNA bands are thus blotted, this is called Southern blotting (after the name of E.M. Southern); when RNA bands are thus transferred, it is described by the jargon term Northern blotting and similarly when protein bands are transferred, the technique is described as Western blotting.

(*a*) *Southern Blotting :* For Southern blotting, DNA sample is first digested with a restriction enzyme and digested sample is gel electrophoresed. The DNA bands in the gel are denatured into single strands with the help of an alkali solution. Subsequently, the gel is laid on top of a buffer saturated filter paper, placed on a solid support (e.g. glass plate), with its two edges immersed in the buffer. A sheet of nitrocelluslose membrane is placed on top of the gel and a stack of many papers (paper towels) on top of this membrane. A weight of about 0.5 kg is placed on top of paper towels. The buffer solution is drawn up by filter paper wick, and passes through the gel to the nitrocellulose membrane and finally to the paper towels. While passing through the gel, the buffer carries with it single-stranded DNA, which binds on to the nitrocellulose membrane, when the buffer passes through it to the paper towels. After leaving this arrangement for a few hours or overnight, paper towels are removed and discarded. The nitrocellulose membrane with single-stranded DNA bands blotted on to it, is baked at 80°C for 2-3 hours to fix the DNA permanently on the membrane. This membrane now has a replica of DNA bands from agrose gel, and can be used for hybridization with radioactively-labelled DNA or RNA probe. The membrane may then be washed to remove any unbound DNA and X-ray film is exposed to the hybridized membrane to get autoradiographs. The above steps involved in Southern blotting are shown in Figure.

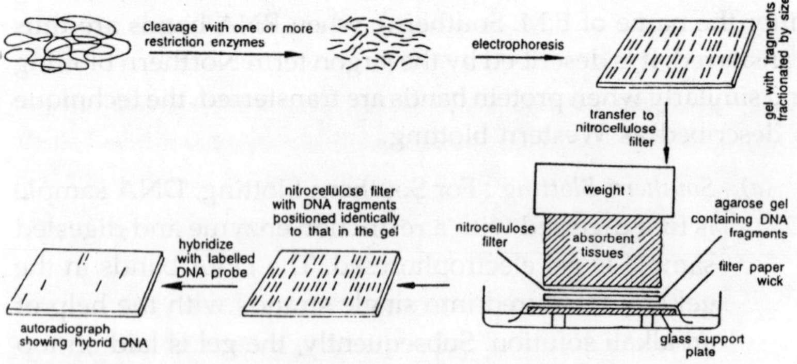

Fig. : Southern blotting method of analyzing DNA segments that share homology with a nucleic acid probe.

(b) *Northern Blotting* : Initially the technique of Southern blotting used for DNA transfer from gel to the membrane could not be used for blot-transfer of RNA. Instead mRNA bands from the gel were blot-transferred onto a chemically reactive paper, prepared by diazotization of aminobenzyloxymethyl paper. The technique being related to Southern blotting was called Northern blotting (not after the name of any scientist as in Southern blotting). Later, it was shown that mRNA bands can be hybridized with a labelled DNA or RNA probe. The single-stranded regions of the hybridized probe are removed by nuclease (mungbean nuclease or S-l nuclease), so that quantitative estimation of hybridized mRNA can also be made.

(c) *Western Blotting* : This technique is used to detect proteins of a particular specificity. Particularly when a transferred gene expresses in transformed cells, the translated product in the form of protein can be identified by this technique. The extracted proteins are subjected to polyacrylamide gel electrophoresis (PAGE) and are then transferred on to nitrocellulose, to which they bind. Nitrocellulose membrane is then used for probing with a specific- labelled antibody (antibody will not hybridize with protein, but bind with it). The antibody may be labelled with ^{125}I and the signal is detected again with autoradiography. If radioactive label is not used, bound antibody may be detected by a second antibody tagged with an enzyme (as in ELISA technique).

Dot blots and slot blots. Another variety of Southern or Northern blots is dot blots or slot blots. In dot blots, cloned or pure extracted DNAs to be tested are spotted adjacent to each other on a nitrocellulose membrane. DNA blots thus produced are immobilized and denatured so that they are bound on the membrane as single-stranded DNA blots. The membrane is then hybridized with readioactively labelled probe (DNA or RNA). The dot representing sequence related to probe will light up by autoradiography (Fig.). The intensity of the dot will indicate the relative concentration of sequence related to the probe in the DNA sample used for a particular dot. No steps involving digestion with enzyme, gel electrophoresis or transfer from gel to membrane are needed in this technique. Therefore, it is considered much more convenient, when no restriction sites are needed to be studied.

Sometimes, an apparatus is used for placing spots on the membranes, through slots made in this equipment. The spots made thus are in the form of oblong slots rather than round dots. These slots are used just like dot blots and are described as slot blots (Fig.).

Molecular Probes Applications

Molecular probes can be used for both basic and applied studies in the field of molecular biology and biotechnology. For instance, molecular probes are extensively used for developing molecular markers, that are used for a variety of purposes.

Building and Screening of Genomic and cDNA Libraries

In order to isolate one or more related genes from a genome, we like to prepare a mixture, of clones each carrying DNA derived either from the genomic DNA or from cDNA (derived from mRNA isolated from a specific metabolically active tissue of an organism). This mixture may contain thousands of clones, which when derived directly from the genomic DNA are collectively called a genomic library. Similarly when these clones are derived from cDNA, they are collectively called a cDNA library. Construction and use of these libraries will be discussed in this section.

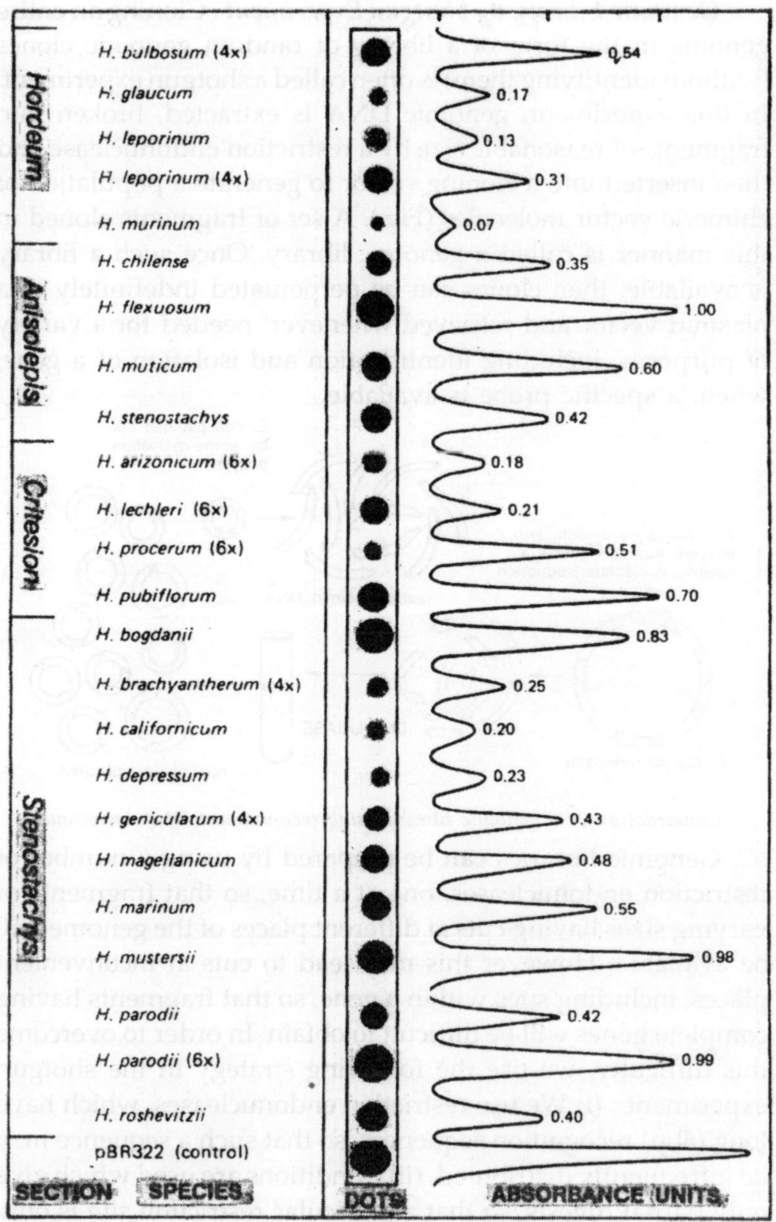

Bot blots form 25 Hordeum species, showing difference in abundance of sequence related to the probe pSC 119.

Genomic Library by Shotgun Experiment: Cloning an entire genome in the form of a library of random genomic clones (without identifying them) is often called a shotgun experiment. In this experiment, genomic DNA is extracted, broken into fragments of reasonable size by a restriction endonuclease and then inserted into a cloning vector to generate a population of chimeric vector molecules (Fig.). A set of fragments cloned in this manner is called a genomic library. Once such a library is available, then clones can be perpetuated indefinitely in a plasmid vector and retrieved whenever' needed for a variety of purposes, including identification and isolation of a gene, when, a specific probe is available.

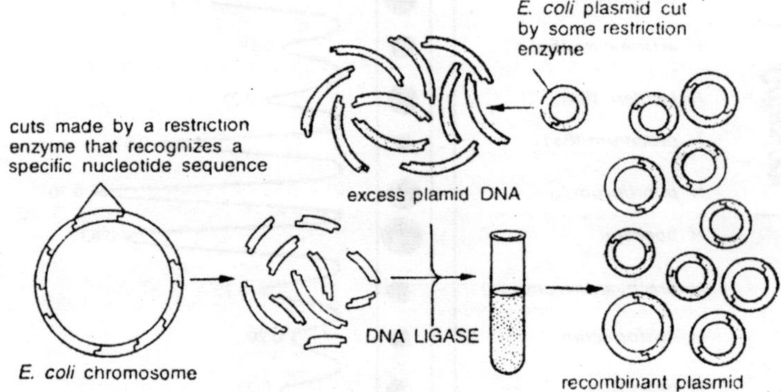

Construction of a genomic library using recombinant DNA technique

Genomic libraries can be prepared by using a number of restriction endonucleases, one at a time, so that fragments of varying sizes having cuts at different places of the genome will be available. However this may lead to cuts at inconvenient places, including sites within a gene, so that fragments having complete genes will be difficult to obtain. In order to overcome this difficulty, we use the following strategy in the shotgun experiment : (i) We use restriction endonucleases, which have long (8bp) recognition sequences, so that such a sequence may be infrequently distributed, (ii) Conditions are used which give only partial digests, so that a particular restriction site is only occasionally cleaved, and long fragments without having any breaks on recognition sites available within a gene can be easily obtained. The number of fragments representing every sequence of the genome increases with genome size. For instance, for a probability level of 99% that all sequences are present in our library of a species, we may need 1,500 cloned fragments for *E. coli*, 4,600 for yeast, 48,000 for *Drosophila melanogaster*, and 8,00,000 for a mammal- like human being. Libraries reaching these desired limits have been prepared in all these cases.

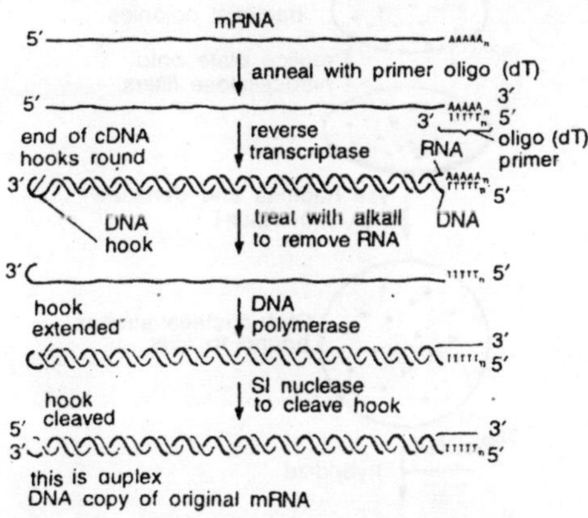

Synthesis of cDNA from mRNA, using reverse transcriptase enzyme.

Library cDNA from mRNAs : Complementary DNA (cDNA) libraries can also be prepared by isolating mRNAs from tissues which are actively synthesizing proteins, like roots and leaves in plants, ovaries or reticulocytes in mammals, etc. The mRNA is used for copying it into cDNA through the use of reverse transcriptase.

A cDNA molecule can be made double stranded, and cloned as shown in Figure. However, cDNA clones will differ from genomic clones in lacking the introns present in split genes, but may have the advantage of being capable to be expressed in bacteria, which do not have the machinery to process HnRNA (obtained from split genes) into mRNA.

Colony (or plaque) Hybridization for Library Screening : Once a genomic library or cDNA library is available, we may like to use it for isolation of a gene sequence. This can be achieved by colony hybridization technique illustrated in Figure. In this technique, bacteria-carrying chimeric vectors are grown into colonies, which are lysed on nitrocellulose filters. Their DNA is denatured *in situ* and fixed on the filter, which is hybridized with a radioactively-labelled probe carrying a

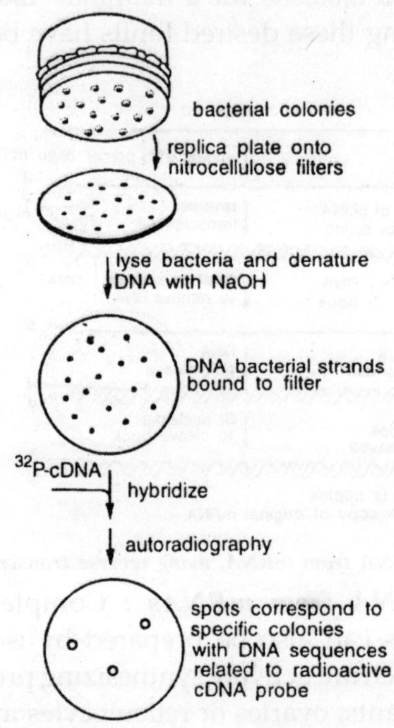

Colony hybridization technique for selection and isolation of a DNA fragment having sequence complementary to a radioactively-labelled probe.

sequence related to the gene to be isolated (usually a cloned cDNA for screening of a genomic library). Colonies carrying this sequence will be identified by dark spots after autoradiography, so that the original chimeric vector carrying the desired gene sequence can be recovered from one or more colonies in the original master plate and used for further experiments. This technique is described as colony hybridization. It is possible that a probe may identify more than one clones or that a gene is fragmented in the library. In such a case, one needs to reconstruct the desired sequence using several overlapping sequences available in the library. This is a very routine exercise whenever we like to isolate specific DNA sequences from the genome of a species, or from cDNA derived from mRNA of a specific tissue of a species.

Sometimes the library may be available, not in the form of bacteria transformed with chimeric DNA molecules, but in the form of chimeric phage particles carrying the cloned segments. In such a situation, a bacterial lawn is infected with a mixture of chimeric phage particles (i.e. the library) and a large number of plaques develop overnight. These plaques can be treated just like the colonies in colony hybridization to identify and isolate the chimeric phage particles carrying the gene of interest. This technique is then described as plaque hybridization.

Walking of Chromosome and Featurisation of Chromosome Segments : When a probe is used for the identification of a gene sequence in a genomic library, the probe may hybridize with a number of clones, each carrying a part of a large gene fragmented during preparation of genomic library. If we obtain partial digests (by digesting the DNA only partially) from the genome, different genomes (form large number of cells) may give fragments which have overlapping sequences, because sites cleaved in different genomes of the same organism will differ being random. Since none of these fragments may have entire sequence represented in the probe, overlapping sequences may be used to construct the original genomic sequence. Identification of fragments with overlapping sequences may be a key to the reconstruction or characterization of large chromosome regions. This is achieved by the technique popularly described as chromosome walking.

The technique of chromosome walking involves the following steps : (i) from the genomic library, a clone of interest (identified by a probe) is selected and a small fragment is subcloned from one end of the clone (there is a technique available to subclone a fragment from the end); (ii) the subcloned fragment of the selected clone may be hybridized with other clones in the library and a second clone hybridizing with the subclone of the first clone is identified due to presence of overlapping region; (iii) the end of the second clone is then subcloned and used for hybridization with other clones to identify a third clone having overlapping region with the subcloned end of the second clone; (iv) the third clone identified as above is also subcloned and hybridized with clones in the same manner and the procedure may be continued; (v) restriction maps of selected overlapping clones may be prepared and compared to know the region of overlapping as shown in Figure, so that identification of few overlapping restriction sites will amount to walking along the chromosome or along a long chromosome segment.

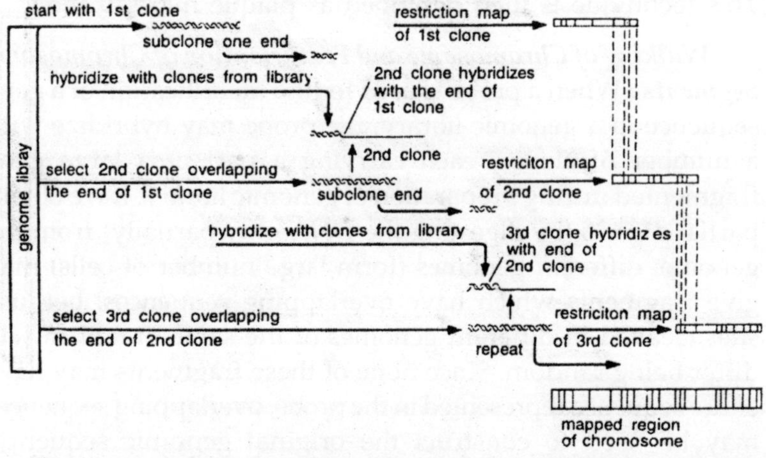

Regions of chromosomes approaching 1000kb have been mapped following the above technique. Restriction maps of entire chromosomes can be prepared in this manner following the technique of chromosome walking.

Characteristics of Chromosome Jumping (or Hopping) Libraries : In most cases, we identify the genes by their protein products, which help in the isolation and cloning of a specific gene. However, for many human disorders, which are each controlled by a single gene (e.g. Huntington disease, cystic fibrosis, duchenne muscular dystrophy, etc.), the gene product is not known despite intensive investigations. Such genes are cloned by a process, in which the gene is identified primarily by its map position. This approach, also described as 'map based cloning' involves the following two steps : (i) locate the gene to a particular chromosome through RFLP linkage analysis; (ii) map the position of the gene with respect to molecular markers, both through recombinational analysis (in terms of recombination value in cM=centiMorgan units) and by physical mapping (in terms of kilobase pairs) with pulsed field gel electrophoresis (PFGE) and preparation of restriction map. This will allow identification of a molecular marker that is closest to the gene. However, even if the distance of the gene from a molecular marker is as small as lcM, which is the limit of resolution, this will be equivalent to 100 to 5,000 kb. Therefore, if a segment carrying the gene is to be cloned using the linked molecular marker, such a segment carrying both the gene and the marker will be too large to be cloned. In order to bring the molecular marker closer to the gene of interest, 'chromosome jumping' approach has been utilized. The technique of

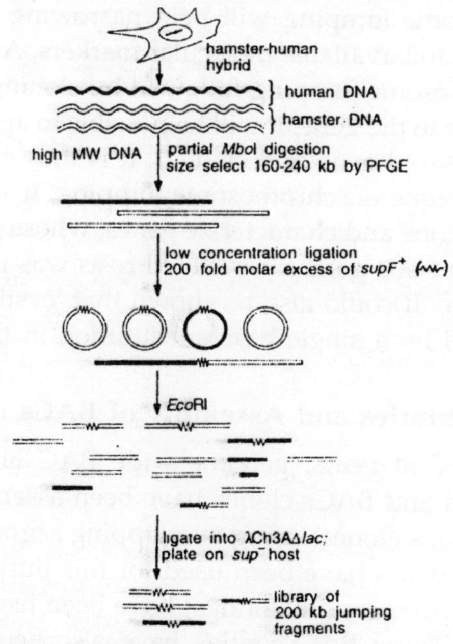

Steps involved in chromosome jumping.

'chromosome jumping' is based on the following steps shown in Figure : (i) depending upon the distance between the gene and the marker, decide about the distance of 'jumps' or 'hopsize' (e.g. 100 kb or 200 kb); (ii) genomic DNA molecules in the range of a particular size (say 80 kb-130 kb in case of 100 kb 'hopsize' or 160-240 kb for 'hopsize' of 200kb) are selected through pulsed-field gel electrophoresis; (iii) for circularization of DNA segments, ligation between two ends of each long linear DNA molecule was allowed using T4 ligase in the presence of *supF*$^{+}$; (iv) DNA circles obtained in step (iii) above are digested with EcoRl; (v) the vector λ Ch3A Δ lac, an amber-mutated phage vector *(supF-)* is also cut with *EcoRl* and used for cloning small DNA fragments representing the junctions of the circularized genomic DNA molecules and carrying *supF*$^{+}$; (vi) the cloned DNA fragments obtained in step (v) above represent the jumping library, which can be plated on a bacterial host and screened through the technique of plaque hybridization described earlier. The above technique of chromosome jumping will help narrowing the gap between the gene and available molecular markers. After several cycles of chromosome jumping followed by cloning the regions that are closer to the gene, it will be possible to approach very close to the desired gene and clone it. It is thus obvious that using the technique of 'chromosome jumping' it will be possible to isolate, clone and characterize genes, whose gene products are unknown. The gene for cystic fibrosis was isolated using this technique. It could also be shown that 'cystic fibrosis' disease is caused by a single base substitution in this gene.

BAC Libraries and Assembly of BACs into Contigs

In recent years, genome wide BAC libraries have been prepared and BACs clones have been assembled into contigs (contiguous clones) using overlapping segments. Fingerprints of these BACs have been used for this purpose and software like FPC (fingerprint contigs) have been used for assembly of contigs. These BAC contigs have also been utilized for the preparation of genome-wide physical maps.

6

Recombinant DNA

Expression Vectors

Techniques for manipulating prokaryotic as well as eukaryotic DNA witnessed a remarkable development during the later part of the 20th century. These techniques involved breakage of a DNA molecule at two desired places to isolate a specific DNA segment and its insertion in another DNA molecule at a desired position.

The product thus obtained is called recombinant DNA and the technique often called genetic engineering. Using this technique we can isolate and clone single copy of a gene or a DNA segment into an indefinite number of copies, all identical. This became possible because vectors like plasmids and phages reproduce in a host (e.g. *E. coli)* in their usual manner even after insertion of foreign DNA, so that the inserted DNA will also replicate faithfully with the parent DNA. This technique is called gene cloning and the vectors used for this purpose are called cloning vectors. With this technique, using a variety of cloning vectors, genes can be isolated, cloned and characterized, so that the technique has led to significant progress in all areas of molecular biology.

A variety of vectors have also been developed which not only allow multiplication, but may also be manipulated in such a way that the inserted gene may express in the host. Due to the importance of a variety of these cloning and expression vectors in genetic engineering experiments, they are discussed in some detail in this chapter. The techniques used for inserting foreign DNA in these vectors and the development of chimeric DNA molecules (for developing molecular probes, gene libraries, etc.).

Scope of Cloning Vectors for Recombinant DNA : One of the most important uses of recombinant DNA technology is the cloning of (i) random DNA or cDNA segments, often used as probes or (ii) specific genes, which may be either isolated from the genome or synthesized organo-chemically or in the form of cDNA from mRNA. This cloning of DNA is possible only with the help of another DNA molecule, which is capable of replicating in a host. This other DNA molecule is often used in the form of a vector, which could be a plasmid, a bacteriophage, a derived cosmid or phagemid (see later in this section), a transposon, a virus or even an artificial chromosome. Techniques should also be available, which will allow selection of chimeric genomes obtained after insertion of foreign DNA from a mixture of chimeric and the original vector (see later). Another critical desired feature of any cloning vector is that it should possess a site at which foreign DNA can be inserted without disrupting any essential function. Therefore, in each case an enzyme will also have to be selected which will cause a single break. Sometimes vectors are modified by inserting a DNA segment to create unique cleavage site(s) for one or more enzymes to facilitate its use in gene cloning. This inserted DNA with restriction sites for several enzymes is sometimes called a polylinker.

Role of Plasmids as Vectors

Plasmids are defined as autonomous elements, whose genomes exist in the cell as extra-chromosomal units. They are self-replicating circular (only rarely linear) duplex DNA molecules, which are maintained in a characteristic number of copies in a bacterial cell, yeast cell or even in organelles found in eukaryotic cells. These plasmids can be single copy plasmids that are maintained as one plasmid DNA per cell or multicopy plasmids, which are maintained as 10-20 genomes per cell. There are also plasmids, which are under relaxed replication control, thus permitting their accumulation in very large numbers (upto 1000 copies per cell). These are the plasmids which are used as cloning vectors due to their increased yield potential.

Circular plasmid DNA, which is used as a vector, can be cleaved at one site with the help of a restriction enzyme to give a linear DNA molecule. A foreign DNA segment can now be inserted, by joining the ends of broken circular DNA to the two ends of foreign DNA, thus regenerating a bigger circular DNA molecule that can now be separated be gel electrophoresis on the basis of its size (Fig.). Selection of bacterial colonies carrying chimeric plasmid DNA is also facilitated by the resistance genes, which the plasmid may carry against one or more antibiotics. If a plasmid has two such genes conferring resistance against two antibiotics and if the foreign DNA insertion site

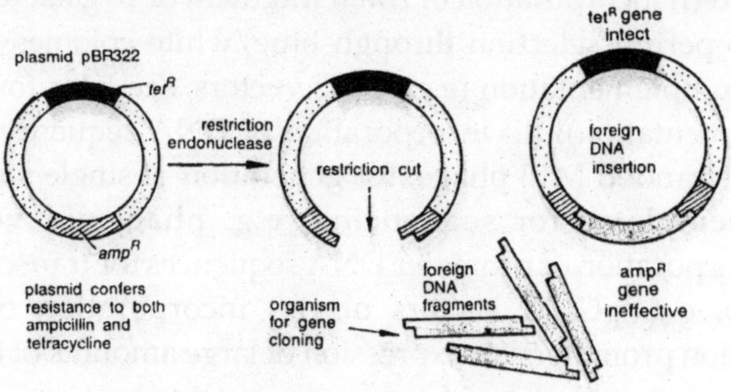

lies within one of these two genes, then the chimeric vector loses resistance against one antibiotic, the gene for which has foreign DNA inserted within its structure. In such a situation, the parent vector in bacterial cells can be selected by resistance against two antibiotics and the chimeric DNA can be selected by retention of resistance against only one of the two antibiotics.

Three Stages of the Formation of Plasmid Cloning Vectors: Development of plasmid vectors proceeded chronologically in the following three phases. (1) In the first phase, the vectors that were made available during 1973-1976 included pSCl0l, ColEl and pCRl, all of them suffering with the following drawbacks : (i) they replicated poorly, (ii) they carried unsuitable selectable markers and (iii) they carried not more than two restriction sites for cloning. These drawbacks were taken care of in pBR313, but it was unnecessarily large, half of its DNA being non-essential. This was reduced in size giving rise to pBR322 (4.362kb ; both pBR313 and pBR322 were produced in 1977), which remained popular for many years. In the second phase, the size of plasmid vectors was greatly reduced, because the efficiency of transformation is inversely related to the size of vector, and also because larger plasmids replicate to lower copy number. A number of derivatives of pBR322 were thus released, which included pAT153, pXf3, pBR327, etc. They made use of antibiotic resistance for selecting recombinant clones. The third phase of the development of plasmid vectors involved (i) incorporation of *Hindi* fragment of β -galactosidase gene to permit selection through blue/white colonies due to alpha-complementation (e.g., pUC vectors; see later for alpha complementation), (ii) incorporation of DNA sequences from single- stranded M13 phage, for generation of single stranded DNA templates for sequencing, e.g. phagemid vectors, (iii) incorporation of promoter DNA sequences for transcription in *vitro,* e.g. pGEM vectors or (iv) incorporation of high expression promoters for expression of large amounts of foreign proteins (e.g. a variety of expression vectors).

Alpha-complementation for Screening, Using "Blue-white" Colonies : The *E. coli lacZ* gene product, β -galactosidase consists of a-fragment at the N-terminal end and ω -fragment at the C-terminal end. The active form of β -galactosidase consists of a tetramer with four identical subunits of 1021 amino acids each. A small deletion in this gene encoding amino acids 11-41 in the region of a peptide (as in *E. coli* strain Δ Ml 5) prevents formation of the tetramer thus leading to loss of β -galactosidase activity. The a-peptide (a small N-terminal fragment consisting of the first 92 amino acids), encoded by the 5' end of *lacZ* gene (present in some vectors, e.g. pUC, M 13), can interact with defective β galactosidase to produce active tetrameric form of the enzyme. This phenomenon is described as α -complementation and is utilized in several vectors for selection of recombinant DNA molecules (phage plaques or the host *E. coli* cells carrying plasmid vector with foreign DNA). Such a vector has the following general features : (i) It contains a *Hindi* fragment of the *lac* operon that includes promoter, operator and a short DNA sequence belonging to the 5' end of *lacZ* gene encoding α -peptide needed for a-complementation. When this vector carrying *Hindi* fragment of *lacZ* gene infects *Lac*⁻ host (E. *coli* Δ M 15), cells within the host bacterial colony or phage plaque will be phenotypically *Lac*⁺ due to a—complementation, (ii) The *Hindi* fragment has been engineered to contain a short DNA sequence which not only encodes a part of α -peptide, but also carries cleavage sites for several restriction enzymes (e.g., M13mpl8 contains EcoRI, *Sstl, Kpnl, Smal, BamHl, Xbal, Sail, Pstl, Sphl, Hindlll* in this order). Insertion of DNA into any of these sites destroys the a—complementing ability of the *Hindi* fragment, giving white colonies/plaques.

Vectors such as pUC and M13mp series contain *Hindi* fragment as above and lack the DNA sequence for co-fragment. These vectors need to be grown on a strain that carries DNA sequence encoding complementary ω -fragment, so that complementation between the two may give β -galactosidase activity. The DNA sequence for ω -fragment is carried by F plasmid in *E. coli* cells (AM 15 strain). When these *E. coli* cells each carrying an α -complementing vector are grown on a medium containing isopropyl thiogalactoside i.e. IPTG (which inactivates the repressor and thus works as an inducer), α - fragment is synthesized, which is complemented by the a-fragment, synthesized due to the 5' segment of the *lac*-Z gene in the vector (Fig.).

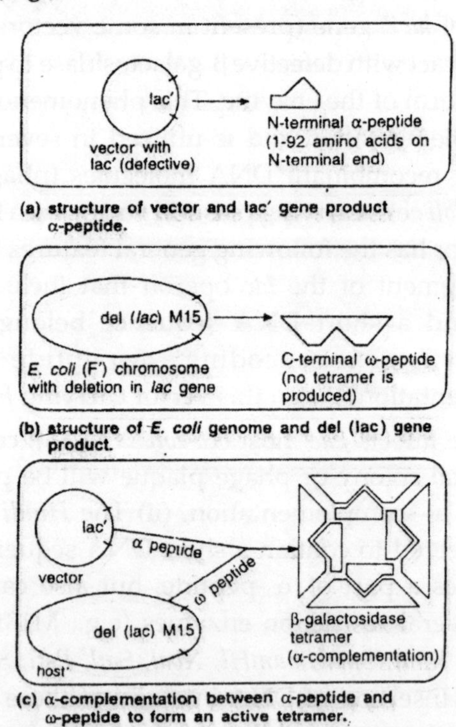

Alpha complementation between α -fragment synthesized due to vector and ω -fragment synthesized due to host.

The α -complementation is detected with the help of a chromogenic substrate called X-gal (5-bromo-4-chloro-3-indolyl-β -D-galactoside), which is a complex between a blue dye

(5-bromo-4 chloroindole) and a molecule-like galactose. This compound is white, but in the Fig. Alpha complementation between α -fragment synthesized due to vector and α -fragment synthesized due to host presence of β -galactosidase (due to α -complementation), when galactose is dissociated, it is rendered blue. However, if a DNA fragment is inserted in the *Lac-Z* gene for a-fragment in the vector, the defective a-peptide synthesized is incapable of bringing about α -complementation and white colonies instead of blue colonies are obtained. This feature allows selection of *E. coli* cells carrying vectors (or M13 plaques) containing foreign DNA within the region of the gene for α -fragment. The development of a simple colour test has greatly simplified the identification of recombinants constructed in plasmid vectors of this type. Using this test, one can screen many thousands of colonies that carry the recombinant plasmids. Recombinant phages (M13) can be similarly selected by visually examining the colour of the plaques. The presence of instert DNA can be verified by restriction analysis of plasmid DNA.

Nonsense Suppressors (in E. coli) to Complement Termination Codons in Vectors : Some vectors (e.g., phage lambda, Charon 4A and Charon 21 A) used in' recombinant DNA research contain nonsense mutations in essential genes. These vectors cannot be propagated in all strains of *E. coli,* but only in some specific strains containing *nonsense suppressor* genes, so that the defective genes in the vectors are translated despite the presence of nonsense mutations (stop codons). For instance, charon 4A and charon 21A need, but charon 32-35 and charon 40 do not need suppressor gene in the host. The nonsense suppressor genes, as we know are tRNA genes, which can read the UAG (amber codon) or UAA (ochre codon). Rarely UGA (opal codon) is also similarly utilized. A number of such nonsense suppressors *(sup1, sup2, sup3, supE, supF)* have been used in developing *E. coli* strains that complement vectors carrying genes with nonsense mutations.

Commonly Used E. Coli Plasmid Vectors.

(a) pBR322 and pBR327 Vectors : The naturally occurring plasmids may not possess all the essential properties of a suitable cloning vector. Therefore, one may have to restructure them by inserting genes of relaxed replication and genes for antibiotic resistance. This has actually been done and suitable plasmid vectors have been obtained. One of the standard cloning vectors widely used in gene cloning experiments is I pBR322 (derived from *E. coli* plasmid ColEl), which is 4,362 bp DNA and was derived by several alterations in earlier cloning vectors (pBR322 was named after Bolivar and Rodriguez, who prepared this vector). It has genes for resistance against two antibiotics (tetracycline and ampicillin), an origin of replication and a variety of restriction sites for cloning of restriction fragments obtained through cleavage with a specific restriction enzyme (Fig.).

Another vector pBR327 was derived from pBR322, by deletion of nucleotides between 1,427 to 2,516. These nucleotides were deleted to reduce the size of the vector and to eliminate sequences that were known to interfere with the expression of the cloned DNA in eukaryotic cells. pBR327 still contains genes for resistance against two antibiotics (tetracycline and ampicillin). In the past, both pBR322 and pBR327 used to be very common plasmid vectors but have now been replaced by more efficient and improved vectors.

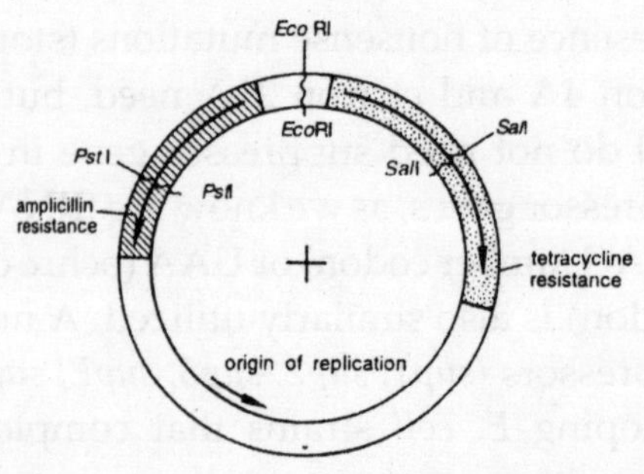

Plasmid pBR 322 and its major characteristic

(b) *pUC Vectors* : Another series of plasmids that are used as cloning vectors belong to pUC series (named after the place of their initial preparation i.e. University of California). These plasmids are 2,700 bp long and possess (i) ampicillin resistance gene, (ii) the origin of replication derived from pBR322 and (iii) a part of *lacZ* gene encoding cc-peptide, so that selection of recombinant vectors through blue/white colonies involving a complementation may be possible. Within the *lacZ* region is also found a polylinker sequence having unique restriction sites. The cloning vectors belonging to pUC family are available in pairs with reversed orders of restriction sites relative to *lacZ* promoter pUC8 and pUC9 make one such pair (Fig.). Other similar pairs include pUC12 and pUC13 or pUC18 and pUC19.

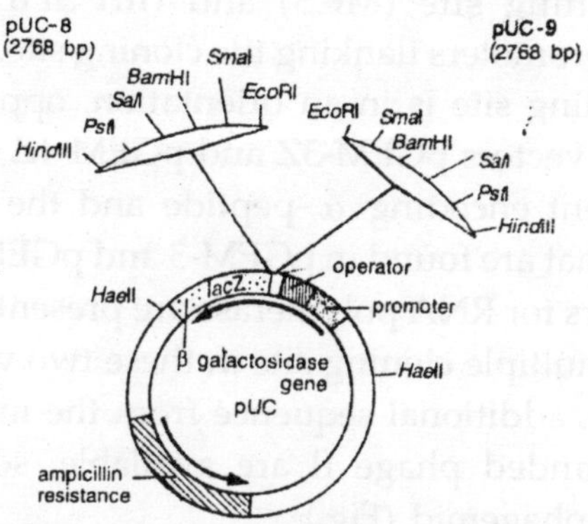

Plasmid pUC and its major characteristic features.

(c) *pSP64, pSP65* : These vectors are derivatives of pUC plasmid vectors and carry promoters for phage RNA polymerase (SP6, T7 or T3) to allow transcription *in vitro* (Fig.). pSP64 and pSP65 contain the phage promoter SP6 and the polycloning site arranged in opposite orientations to allow transcription of the desired strand from the cloned double stranded DNA. Derivatives of pSP64 and pSP65 i.e. pSP64CS and pSP65CS: have been designed to facilitate sequencing by Maxam-Gilbert's method.

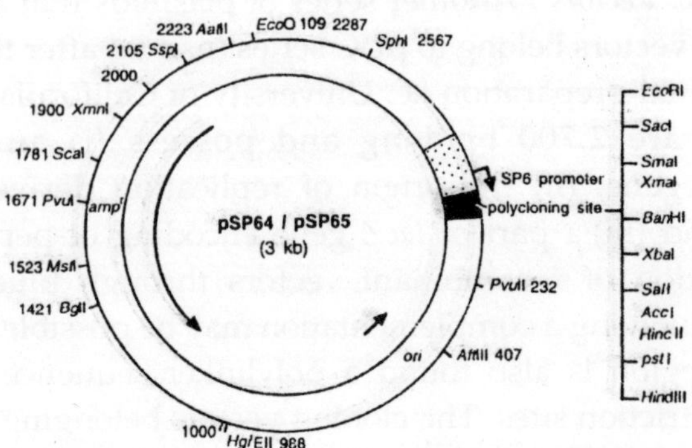

Plasmid pSP64 and its characteristic features.

(d) *pGEM-3, pGEM-4, pGEMa-32, pGEM-4Z* : The vectors pGEM3 and pGEM4 carry (i) ampicillin resistance gene, (ii) a multiple cloning site (MCS) and (iii) SP6 and T7 RNA polymerase promoters flanking the cloning site. In pGEM4, the multiple cloning site is in an orientation, opposite to that in pGEM3. The vectors pGEM-3Z and pGEM-4Z carry *lacZ* gene *Hindi* fragment encoding α -peptide and the same multiple cloning site that are found in pGEM-3 and pGEM-4, but T7 and SP6 promoters for RNA polymerase are present at the opposite ends of the multiple cloning site in these two vectors (Fig.). In pGEM-3Zf(-), additional sequence from the intergenic region of single stranded phage fl are available, so that this will behave as a phagemid (Fig.).

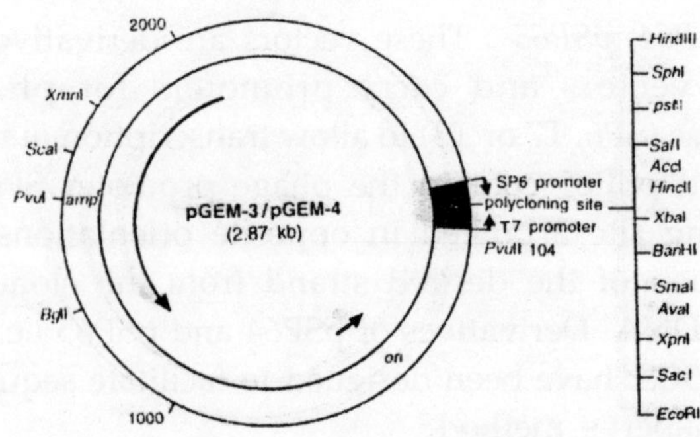

Plasmid pGEM-3 and its characteristic features.

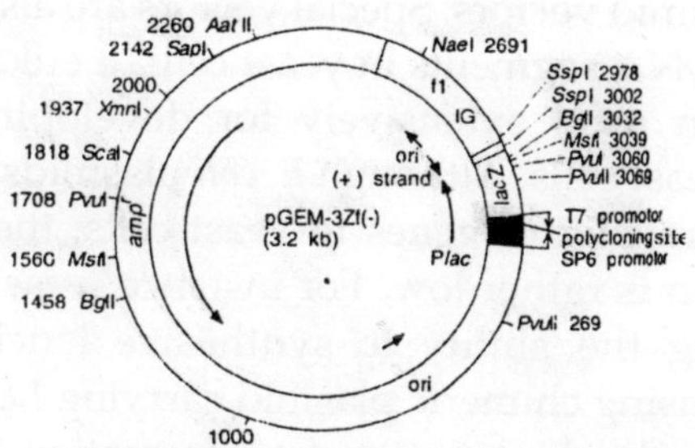

Plasmid pGEM-3Zf(-) and its characteristic features.

(e) *vAN 13 (drived from nAN7).* 7T.AN13 was derived from 7T.AN7, which in its turn was derived from TtVX. It contains a pBR322 origin of replication, pUC polycloning site and a *supF* tRNA genc (Fig.). 7T.AN13 has been used for recombination screening. Probe sequence is cloned in the 7T.AN13 plasmid vector carrying *supF* and the recombinant plasmid is introduced into *E. coli.* Libraries constructed in phage A, vectors that carry several amber mutations in essential genes are also grown in bacteria containing recombinant 7T.AN13 plasmid. If there are homologous sequences between plasmid and *X* vectors, the whole plasmid is transferred to *X* vectors by recombination, so that in the cointegrants, *supF* of 7iAN13 suppresses amber mutation in one or more essential genes carried by the phage thus permitting plaque formation in non-suppressing hosts. Thus the genes carrying sequences homologous to the probe carried by π AN13 (e.g. multigene families, genomic sequence corresponding to cDNA, etc.) can be isolated.

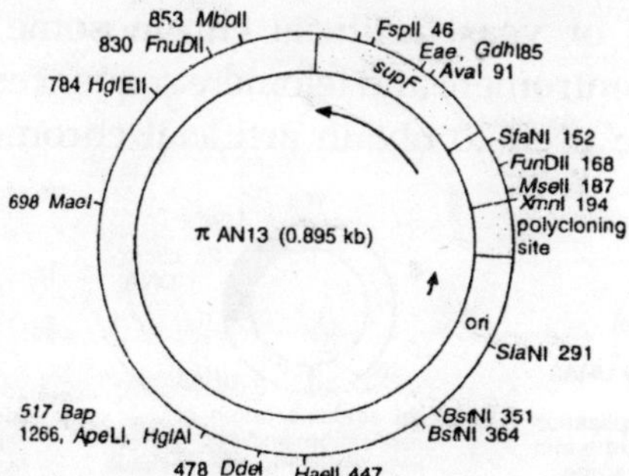

Plasmid π AN13 (carrying supF) and its characteristic features.

Yeast plasmid vectors. Special vectors are also available for introducing DNA segments in yeast cells, a eukaryotic system that has been used extensively for developing genetically engineered yeast cells. Although *E. coli* plasmids or phages can be used for transfer of genes to yeast cells, the frequency of transformation is rather low. For instance, yeast cells that are *LELT* (lacking the ability to synthesize leucine) could be transformed using chimeric plasmid carrying *LEU+* gene from yeast. This involved integrative transformation (crossing over) having low frequency. To overcome this difficulty of low frequency transformation, efforts were made to prepare a variety of vectors for yeast (Fig.).

The possible examples of the yeast plasmid vectors are :

(i) Yip or yeast integrative plasmids, which allow transformation by crossing over and have no replication origin;

(ii) YEp or yeast episomal plasmids, which carry 2 μ DNA sequence influencing the origin of replication and *rep* genes;

(iii) YRp or yeast replicating plasmids, which carry autonomously replicating sequence; this sequence is very common with many yeast genes so that stable transformation can be achieved by crossing over;

(iv) YCp or yeast centromere (CEN) containing plasmids, which function as true chromosomes and segregate during mitosis and meiosis;

(v) pYAC or yeast artificial chromosome, which carries both centromere and telomere sequences and, therefore, can be used to obtain artificial chromosomes.

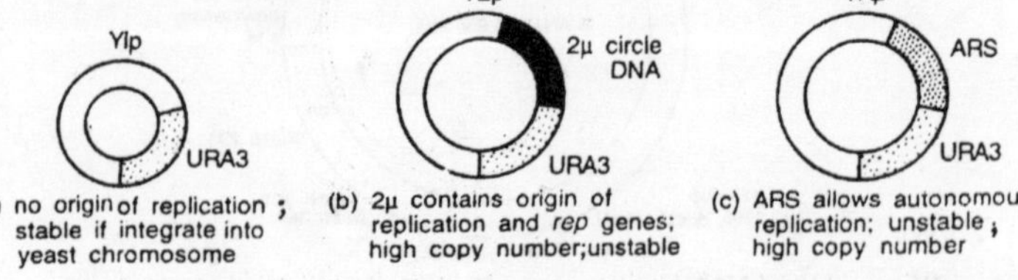

(a) no origin of replication ; stable if integrate into yeast chromosome

(b) 2μ contains origin of replication and *rep* genes; high copy number;unstable

(c) ARS allows autonomous replication; unstable ; high copy number

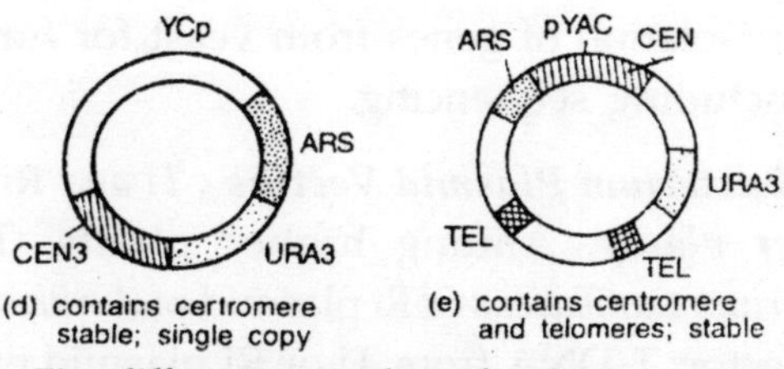

Five different yeast plasmid vectors.

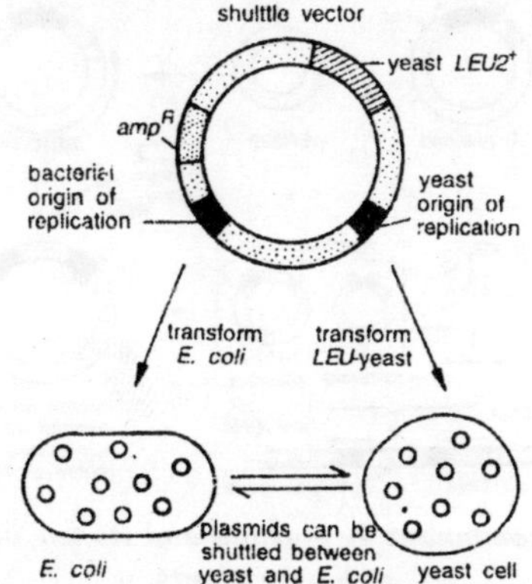

A shuttle vector having both yeast and bacterial origins of replications.

The above vectors have been extensively utilized for a study of yeast genome and are also used as shuttle vectors, which allow genie sequences to be routinely transferred back and forth between yeast and *E. coli* cells, provided they contain origins of replication that are active in both yeast and *E. coli.* These vectors also have markers enabling the selection of *E. coli* cells or colonies transformed with these vectors. The vectors can, therefore, be amplified in *E. coli* and then used to transform yeast cells (Fig.).

Retriever vectors are another class of plasmid vectors, which are used to retrieve specific genes from the normal chromosome of an organism like yeast through recombination. They will multiply both in *E. coli* and yeast. In such a vector the gene in question is removed by enzymes to produce gapped vector having flanking sequences. Through recombination, they acquire lost sequences from the host. Retriever vectors are very useful for isolation of genes from yeast for further molecular studies including sequencing.

***Agrobacterium* Plasmid Vectors** : *Ti and Ri plasmid vectors for higher plants.* Among higher plants, Ti plasmid of *Agrobacterium tumifaciens* or Ri plasmid of *A. rhizogenes* is the best known vector. T-DNA from Ti or Ri plasmid of *Agrobacterium*

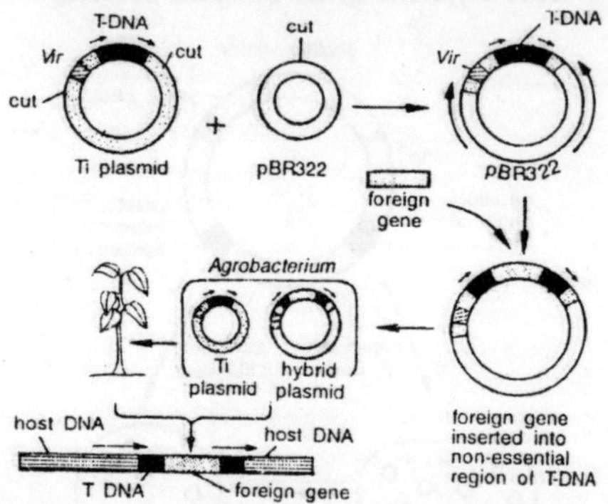

Derivation of recombinant Ti plasmid and its use in transformation of higher plants.

is considered to be a very potential vector for cloning experiments with higher plants. This will involve the following steps : (i) foreign DNA has to be first cloned into T-DNA of Ti or Ri plasmid; (ii) modified hybrid T-DNA can be transferred to the genome of plant cells by *Agrobacterium* infection. (Fig.). This system can be widely used, since *Agrobacterium* infects nearly all dicotyledonous plants. Such cloning in plants has proved to be of immense use to modify agricultural plants to increase their productivity. Monocotyledonous plants to which our cereal crops belong are, however, difficult materials to be used for improvement through *Agrobacterium* mediated transformation; success, however, has been achieved in cereals also in recent years.

Role of Lambda Phage (k) Vectors : Plasmid vectors described before are often used for cloning DNA segments of small size (upto 10 kilobases). However, while preparing a genomic library for a eukaryote, the cloned fragments should be large enough to contain a whole gene. This will also allow cloning of the whole genome into a number of clones, which will not be unreasonably large and therefore can be screened without serious difficulty.

The above properties and other requirements of cloning whole genome in eukaryotes are fulfilled by the phage lambda *(X)* and cosmid vectors, the former permitting cloning of segments upto 20-25kb long (kb = kilobases) and the latter accommodating segments upto 45kb long. Phage lambda *(X)*, however, is easier and more efficient for making genomic and cDNA libraries.

Lambda Phage (Lysis and Lysogeny) : The λ phage has a genome consisting of 50kbp circular DNA. It follows either a lytic path (large number of phage particles are released due to lysis) or a lysogenic path (phage genome integrates with host's genome due to expression of gene *cl* and then replicates with host's genome). Lytic path may be switched towards lysogeny and *vice versa* (Fig.). For an account of genetic regulation of lytic and lysogenic pathways, readers are advised to consult author's text book *"Genetics"*.

An important feature of lambda *(* λ *)* genome is that only about 60% of the genome (including left arm consisting of genes A-J measuring 20kb and the right arm, containing genes O-R and measuring 8-10kb in length) is essential for lytic growth of the phage and the central portion (40%) can be replaced by foreign DNA. The viability of phages goes down drastically, when its genome size is outside the acceptable limit of 78%-105% of the wild genome. This fixes the upper as well as the lower limits of size of foreign DNA which can be cloned.

A single break in the linear λ phage DNA will generate two fragments, which are later joined together with foreign DNA to generate a chimeric phage particle (Fig.). The chimeric phage can be multiplied after allowing it to infect bacteria and collecting progeny particles after a lytic cycle. The use of phage particles as vector imposes a limitation on the size of foreign

DNA, which can be cloned, because the capacity of phage head is only limited, and if the size of foreign DNA is too long, size of phage DNA may not be accommodated in phage head. In order to overcome this problem, those segments of DNA, which do not contain essential genes, may be removed. Such a technique has been followed in phage lambda *(X)* to create a smaller vector genome having single restriction site for the enzyme *EcoRI*. Since the reduced size also fails to be adequately packed in phage head (there is also a requirement of a minimum size of DNA), this automatically provides a selection method, in which only the chimeric particles will be obtained in the phage progeny, and the vector particles lacking a cloned segment will be eliminated due to their reduced size.

Genetic Markers for the Selection and Screening of λ Vectors

(a) β Galactosidase : Some of the λ vectors contain a segment of *E. coli lacZ* gene coding for a peptide, that is capable of complementing the defective *lacZ* gene (deletion in the region of α peptide) in the host giving blue colonies. These vectors give white colonies, due to lack of a complementation, whenever there is an insertion of foreign DNA.

(b) *Lysogeny* : Some of the vectors contain a *cl* gene with an *EcoRI* site, so that the vector is *Ct*$^{+}$ and forms lysogens in hosts carrying a mutation *hfl* (high frequency lysogenization). Lytic growth is greatly reduced, when wild type vector with intact *cl* infects, but *cl* gene is disrupted, when foreign DNA is inserted, leading to lytic growth and phaque formation.

(c) *Sensitivity to P2 Interference (Spi*$^{+}$ *Spf*$^{-}$*)* : In some of the replacement vectors, genes *red* and *gam* (genes involved in recombination) are present in the stuffer region, rendering them Spi^{+} (sensitive to P2 prophage interference, i.e. restricted growth on the host, which are lysogens and carry P2). When stuffer region is removed, *red* and *gam* are lost rendering the recombinant vector *red*$^{-}$ *gam*$^{-}$, and consequently Spi^{-} (growing well in P2 lysogens as long as they carry *chi* site and *rec*$^{+}$). The vectors carrying *red* and *gam* in the stuffer region include λ 2001, λ DASH and EMBL series.

(d) Chi Site : Wild type lambda contains no *chi* site (a suppressor of small plaque phenotype in *red⁻ gam*) and therefore when they are *red⁻ gam~~* genotype, in the presence of *chi* site, it does not form small plaques, but instead forms normal plaques as in case of *red+ gam⁺* genotype. It is for this reason that in the replacement vectors like those of EMBL series, *red* and *gam* genes, are present in the stuffer region and *chi* site is present elsewhere, so that when *red* and *gam* are lost with the stuffer, the vector, despite being *red⁻ gam⁻* forms large colonies (due to *chi* site).

Growth of Lambda (λ) Phage Vectors : There are two classes of lambda cloning vectors : i (i) insertion vectors, which accept inserts, upto 12kb long at a single multiple cloning site (MCS) (e.g. λ gt vectors and λ ZAP vectors); and (ii) replacement vectors, which accept inserts, 9-23kb long, involving replacement of a non-essential part (stuffer) of genome (e.g. λ FIX vector; λ DASH vector, EMBL series; Charon series); they have a pair of sites flanking a segment of nonessential DNA that can be cut, removed and replaced by foreign DNA.

(a) Insertion λ Vectors, (i) λ gt10 and λ gt11. λ gt10 and Xgt11 are modified lambda phages designed to clone cDNA fragments. The major difference between these two vectors is that λ gt11 is an expression vector, where inserted DNA is expressed as β galactosidase fusion protein (Fig.). λ gt10 is a 43kb double stranded DNA for cloning fragments that are only 7kb in length. The insertion of DNA inactivates *cI⁺* (repressor) gene generating a *cT* bacteriophage (*cF* phage is unable to lysogenize). Non-recombinant *λ gt10* is *cI⁺* and forms cloudy plaques due to lysogeny on appropriate *E. coli* host, while recombinant *cI⁻ λ gt10* forms clear plaques due to lysis permitting screening of recombinant plaques. Further, in an *E. coli* strain carrying *hflAl50* mutation (high frequency lysogeny mutation) only *cF* phage will form plaques, because *cI⁺* will

form lysogens (integrate with bacterial genome) and will not undergo lysis to form any plaques. Recombinant λ *gt10* plaques can thus be i easily selected. λ *gt*11 is a 43.7kb double stranded λ phage for cloning DNA segments, which are less than 6kb in length (usually for cDNA). Foreign DNA can be expressed as P galactosidase fusion proteins. Recombinant λ *gt*11 can be screened using either nucleic acid or antibody probes (λ *gt10* can not be screened using antibodies). The recombinant λ *gt11* becomes *gar,* while non-recombinant λ *gt11* remains *gat,* so that an appropriate *E. coli* host, with recombinant phage *(gat)* will form white or clear colonies and that with non-recombinant phage *(gat)* will form blue colonies permitting screening in the presence of IPTG (inducer) and λ gal (substrate).

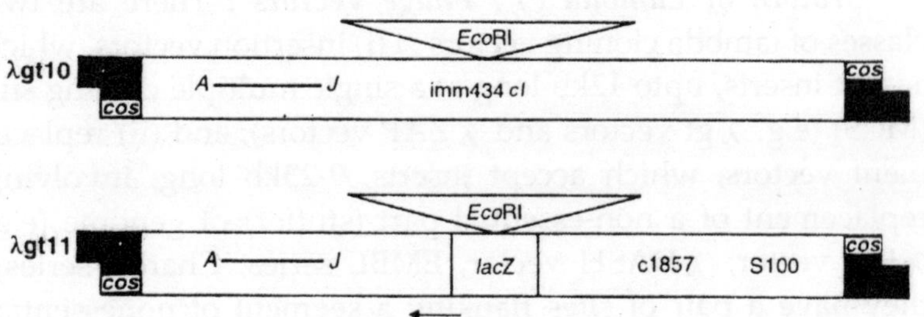

Insertion lambda (λ) vectors λ gt10 and λ gt11 and their essential elements.

(ii) Lambda (λ) Zap Vectors : These vectors (designed by Stratagene) combine the high efficiency for the construction of recombinant lambda and the convenience of a plasmid system with blue-white colour selection. λ ZAP vectors also contain complete sequence of phagemid vector pBluescript, which can be excised *in vivo.* These vectors can accommodate DNA inserts, upto 10kb in length and have the following additional features (Fig.) : (i) they can be screened with either the DNA probes or antibody probes; (note : selection and screening are two steps in the construction and use of genomic libraries; selection allows identification of colonies or plaques with recombinant vector molecules; screening allows identification of specific recombinant vector of one's own choice);

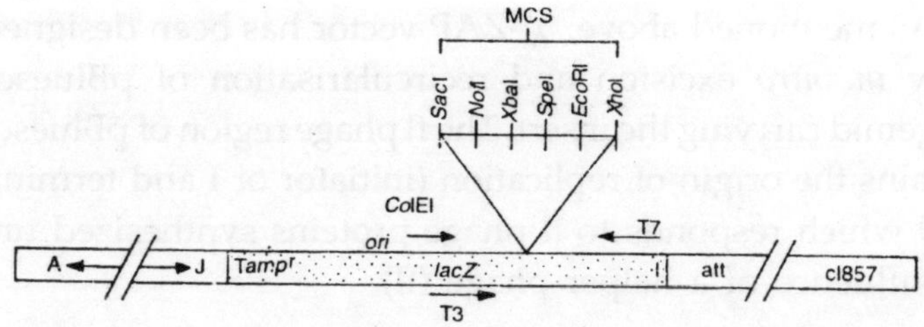

(ii) they allow *in vivo* excision of the pBluescript phagemid, thus permitting characterization of the insert in a plasmid system (Fig.); pBluescript also has the becteriophage fl (single stranded DNA phage) origin of replication (initiation and termination sites) permitting rescue of single stranded DNA, which can be used for DNA sequencing or site directed mutagenesis (exonuclease III and mungbean nuclease can be used for creating deletions at either end); (iii) the polylinker of pBluescript present in λ ZAP vector provides a multiple cloning site (MCS), flanked by T3 and T7 promoters (these can be used to produce riboprobes or for PCR reactions), and a number of primer sites (T3, T7, Ml3 forward, M13 reverse, SK, KS) for DNA sequencing, (iv) they also have *lacZ* promoter to allow expression of fusion proteins (screened through antibodies) suitable for Western blots or protein purification.

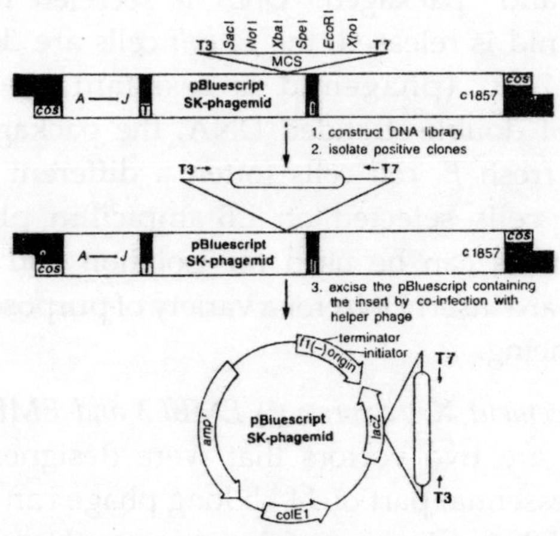

Excision of plasmid pBluescript with the insert form λ Zap vector.

As mentioned above, λ-ZAP vector has been designed to allow *in vitro* excision and recircularisation of pBluescript phagemid carrying the insert. The fl phage region of pBluescript contains the origin of replication (initiator or I and terminator or *T)* which responds to fl phage proteins synthesized under the influence of a helper phage (fl).

Therefore, for *in vivo* excision of pBluescript, *E. coli* cells are simultaneously infected by lambda vector and fl helper page. The fl phage proteins nick one of the two DNA strands at initiator site and allow replication of DNA downstream (upto the terminator T site), thus duplicating complete pBluescript carrying the insert.

The single-stranded DNA thus synthesized is circularized by gene II product of fl phage, so that a circular DNA molecule containing all sequences between initiator and terminator sites is produced (this will include entire pBluescript with the insert, if any; it will not include sequences of lambda phage located outside; Fig.).

Circularization also recreates fl origin. Signals for packaging of fl phage are found at the terminator (T) site, so that the circularized and "packaged" DNA is secreted from *E. coli.* Once phagemid is released, the *E. coli* cells are destroyed by heating to 70°C (phagemid is resistant to 70°C). For production of double-stranded DNA, the packaged DNA is mixed with fresh *E. coli* cells (often a different strain) and transformed cells selected on LB-ampicillin plates. These selected colonies can be used for isolation and analysis of double standard insert DNA for a variety of purposes including DNA sequencing.

(b) Replacement X Vectors : (i) *EMBL3 and EMBL4.* EMBL3 and EMBL4 are two vectors that were designed so that a central non-essential part of 44 kb long phage can be replaced by a foreign DNA. Cleavage of the phage with an appropriate enzyme generates three fragments (left arm, right arm and a

central fragment called stuffer). The central fragment representing 40% of the phage genome is non-essential for propagation of the phage and can be replaced by foreign DNA that may be as long as 20-23 kb.

Non-recombinant DNA resulting from fusion of left arm and right arm will give a DNA molecule that is too small for packaging, so that there is automatic selection against non-recombinant phages (Fig.). EMBL3 and EMBL4 can be used for preparing genomic libraries in eukaryotes with cloned fragments, 15-25 kb in size. The two vectors have polylinkers with reverse orders of restriction sites with respect to each other (Fig.).

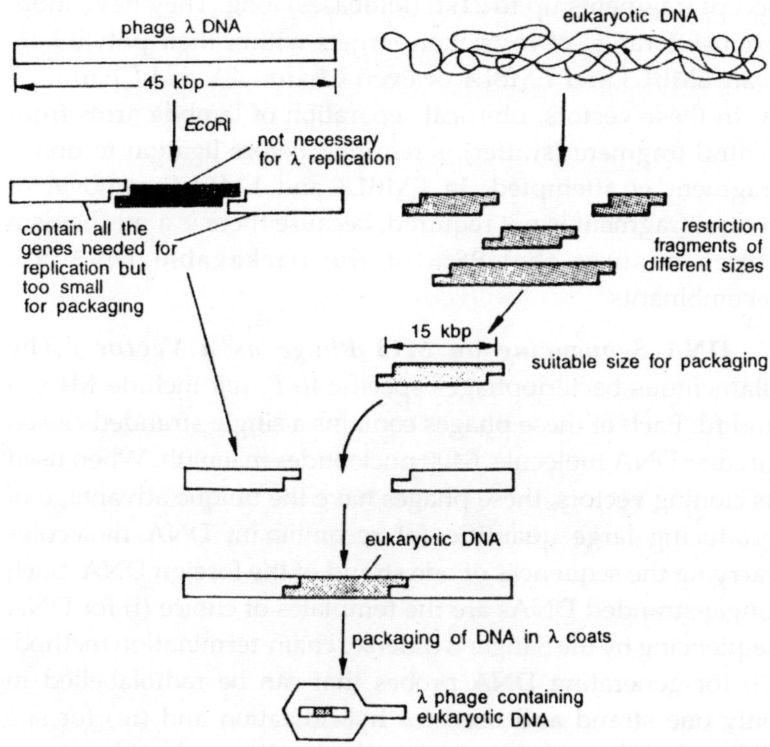

Use of lambda (λ) phage as replacement vector for cloning a eukaryotic DNA fragment.

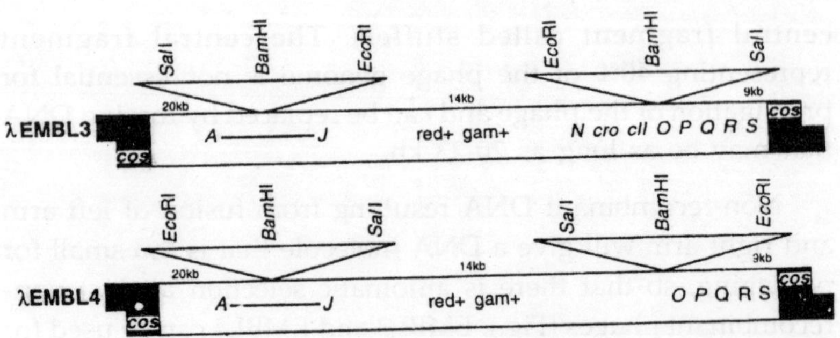

Replacement lambda vectors EMBL3 and EMBL4 showing their essential elements

(ii) *Charon 34 and Charon 35* : Charon 34 and Charon 35 differ from each other only in their central fragments and will accept fragments up to 21kb (kilobases) long. They have more extensive range of restriction targets within their polylinkers than EMBL3 and EMBL4 or even Charon 4A and Charon 21 A. In these vectors, physical separation of lambda arms from central fragment (stuffer) is required before ligation to donor fragment is attempted. In EMBL3 and EMBL4, removal of central fragment is not required, because there is a mechanism which ensures that 98% of the packagable DNA are recombinants.

***DNA Sequencing by M13 Phage as a Vector* :** The filamentous bacteriophages specific to *E. coli* include M13, fl and fd. Each of these phages contains a single-stranded closed circular DNA molecule, 6400 nucleotides in length. When used as cloning vectors, these phages have the unique advantage of producing large quantities of recombinant DNA molecules carrying the sequences of one strand of the foreign DNA. Such single-stranded DNAs are the templates of choice (i) for DNA sequencing by the Sanger's dideoxy chain-termination method; (ii) for generating DNA probes that can be radiolabelled in only one strand and used for hybridization and (iii) for site directed mutagenesis using synthetic oligonucleotides.

M 13, fl and fd may be regarded as identical for purposes

of their use as vectors, although only Ml3 has been mainly used for designing vectors. The infecting single-stranded DNA is designated as the + (plus) strand, which is converted into replicative double- stranded form (RF), in which - (minus) strand is used as the coding strand and for synthesis of + (plus) strands for multiplication of the phage. The M13 genome consists of ten genes (I-X), all of them essential for infection and multiplication of the phage. There are two intergenic IG regions (between genes II and IV and between genes III and VIII), containing sequences for regulating gene expression and DNA synthesis (Fig.), so that these sequences are also essential and not dispensable. Therefore, no sequences can be deleted in designing the vectors. However, in the 500 nucleotides long intergenic region between genes II and IV, foreign DNA can be inserted (this IG region contains signals for packaging and orientation of phage DNA). Insertion of foreign DNA can affect replication of phage genome, but accumulation of mutations in the adjoining genes III or V compensate for this and allow replication. All M13 vectors contain such mutations.

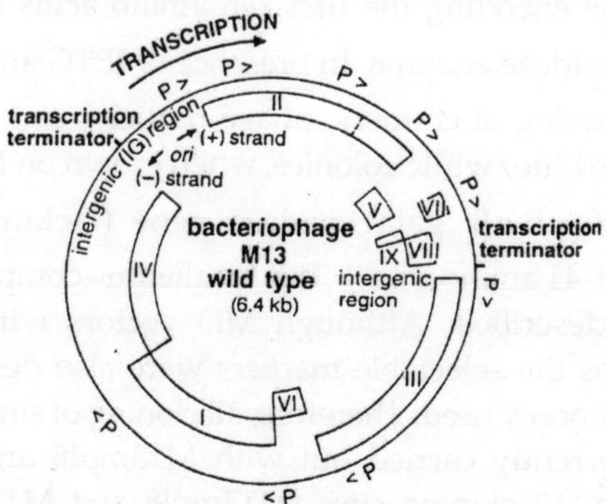

Genome of single stranded M13 phage, showing 10 genes and the intergenic (IG) regions.

Single-stranded DNA is generally a poor substrate for restriction enzymes and for ligases needed for inserting a foreign DNA fragment. Therefore, replicative double-stranded form of the phage DNA is isolated from infected cells, purified and used like plasmid DNA. After insertion of foreign DNA, the chimeric RF is reintroduced into bacterial cells, where it undergoes the normal replication cycle thus generating single-stranded progeny phage particles containing the + strand of the phage along with one of the two strands of the inserted foreign DNA. The other strand (-) of the phage is never packaged.

M13 Vectors of mp Series : As described earlier, the Ml3 genome contains two intergenic regions (one between genes II and IV, the other between genes III and VIII). M13 mp series of vectors make use of intergenic region between the genes II and IV for insertion of foreign DNA. The other smaller intergenic region between the genes III and VIII has been used as the cloning site in another series of vectors, that are less frequently used.

The mp series vectors of Ml 3 were derived from the phage M13mpl, which carries in its major IG region a DNA segment (from *E. coli)* containing some regulatory sequences and a part of *lacZ* gene encoding the first 146 amino acids (α -peptide) of β galactosidase enzyme. In presence of IPTG and X-gal, this allows screening of chimeric phage particles carrying foreign DNA due to blue/white colonies, when grown on *E. coli* strains with a defective β galactosidase gene (lacking sequence encoding 11-41 amino acids). This is called α -complementation as earlier described. Although Ml3 vectors with antibiotic resistance as the selectable markers were also designed, they are not commonly used. Therefore, all cloning of single stranded DNA is currently carried out with M13mpl8 and M13mpl9 which carry 13 cloning sites (M13mpl8 and M13mpl9 carry these sites in opposite orientation) for accepting foreign DNA. Due to opposite orientations of cloning sites in M13mpl8 and M13mpl9, the progeny of one of these vectors will contain one strand of foreign DNA and the progeny of the second vector will contain the complementary strand automatically.

Consequently, by using M13mpl8 and M13mpl9 as dual vectors, it is possible to use single primer (universal primer) to determine the sequence of nucleotides on opposite strands from each end of the inserted DNA. The polycloning sites and their orientations, in M13mpl8 and M13mpl9 are the same as those in pUC18 and pUC19, so that the fragments of foreign DNA can be moved between M13mp vectors and pUC vectors with great ease. Other similar pairs of vectors of mp series are M13mp8 and M13mp9; M13mpl2 and M13mpl3. They correspond to similarly numbered plasmid vectors of pUC series. (pUC8 and pUC9; pUC12 and pUC13), particularly with respect to polylinker cloning site.

Hosts for M13 Vectors. For M13 vectors, a number of host strains have been recommended.

Since M13 enters the host through sex pili, encoded by F factor, only male bacteria carrying Fl episomes are used. These hosts carry a number of genetic markers including the following: (i) *lacZAM15,* a deletion in *lacZ* gene sequence coding for - NH2 terminal region of β galactosidase; this defective *lacZ* gene is sometimes carried on Fl episome; (ii) Δ *(lac⁻ pro-AB),* a deletion causing loss of ability to use lactose and needing proline for growth; (iii) *lacP* causing overproduction of lac repressor, which minimizes the production of toxic proteins encoded by foreign DNA under the influence of lac promoter; (iv) *supE,* a suppressor mutation that allows the growth of M13 vector carrying an amber mutation (M13 vectors were earlier but not now required to carry an amber mutation as a bio-safety guideline).

Cosmids as Vectors

Cosmids are plasmid particles, into which certain specific DNA sequences, namely those for *cos* sites are inserted. Since these *cos* sites enable the DNA to get packed in lambda particle, cosmids allow the packaging of DNA in phage particle *in vitro,* thus permitting their purification. Like plasmids, cosmids also perpetuate in bacteria and do not carry the genes for lytic development. The advantage of the use of cosmids for cloning is that its efficiency is high enough to produce a complete genomic library of 10^6 - 10^7 clones from a mere 1 μ.g of insert DNA. The disadvantage, however, is its inability to accept more than 40-50 kbp of DNA. Bacteriophage PI system and F-factor based vectors (BACs = bacterial artificial chromosomes) can allow cloning of DNA segments, as large as 100 to 1000 kbp (or 1 Mbp = 10^6bp) length. These are, therefore, preferred when fragments bigger than 50-100 kb are to be cloned (see later for details).

Phagemids as Vectors

Phagemids combine desirable features of both plasmids and filamentous bacteriophages. They contain (i) origin of replication from plasmid ColEl, (ii) selectable marker for antibiotic resistance and (iii) major intergenic (IG) region of a filamentous phage, necessary for initiation and termination of

Different phagemid vectors with other details.

	Phagemid	*IG from*	*Helper virus*	*Host E. coli*
1.	pEMBL (derived from pUC8)	fl	IR1 (a variant of fl)	71/18
2.	pBR322	M13	Wild type M13	KK 2186
3.	pRSA101 (derived from π VX)	M13	Avariant of M13	XS127 XS101
4.	pUC 118/119 (derived from pUC 18/19)	M13	M13K07 (with mutant gene II)	MVII84
5.	pBluescript	fl	M13K07	XL1-Blue

viral DNA synthesis and for morphogenesis of phage particles. Segments of foreign DNA cloned in these vectors are multiplied as in plasmid vectors, but when cells harbouring these plasmids are infected with a suitable filamentous phage (helper phage), the phagemid genome behaves like that of a phage. The gene II of the infecting helper phage interacts with the intergenic region of phage available in phagemid, so that the phagemid replicates like filamentous phage and generates multiple copies of one strand of the phagemid and the associated foreign DNA inserted in it. These single stranded-copies of the phagemid DNA (including cloned fragment) will be nicked, circularized and packaged into progeny phage particles. The single-stranded DNA can be purified from these phage particles and used for (i) sequencing, (ii) site directed mutagenesis and (iii) synthesis of strand-specific probes. Phagemids have been developed by inserting into plasmids, the intergenic regions derived from a variety of filamentous phages (Table).

Phagemids have several advantages including the following: (i) like plasmid vectors, they give high yield of double- stranded foreign DNA, (ii) they circumvent the difficulty of subcloning of DNA fragments from plasmid vectors into M13 vectors for sequencing, because single-stranded DNA can be directly obtained using a helper phage, (iii) since they are small in size, single-stranded DNA of upto 10kb can be obtained.

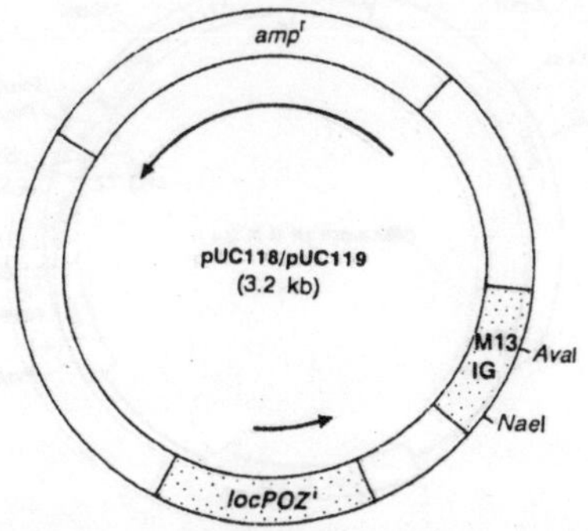

Phagemid versions of plasmids pUC18 and pUC19.

Phagemids pUC118 and pUC119: This is a pair of phagemid vectors derived from pUC18 and pUC19 respectively and contain *cis* acting sequences for (i) initiation and termination of phage Ml3 DNA synthesis and for (ii) packaging of DNA into phage particles (Fig.). These vectors accept segments of foreign DNA and can be propagated as plasmids in a conventional way. When cells harbouring these plasmids are infected with a suitable filamentous phage, copies of only one strand of the plasmid DNA are synthesized and packaged into progeny phage particles. Single-stranded DNA can be isolated from the phage particles and used for a variety of purposes including DNA sequencing.

Phagemid pBluescript IIKS. pBluescript II KS is the most commonly used vector in molecular biology laboratories world over. It is derived from pUC19 and is 2961 bp (bp = base pairs) long. The KS designation indicates the orientation of polylinker, such that the transcription of *lac*Z gene proceeds from the restriction site for *Kpnl* to that for *Sad*. It is also described as pBluesrcipt Ml3 + / -, because it carries M13 origin of replication in opposite orientations to generate single-stranded DNA *in vivo* or RNA in *vitro*. It has been derived from pUC plasmid

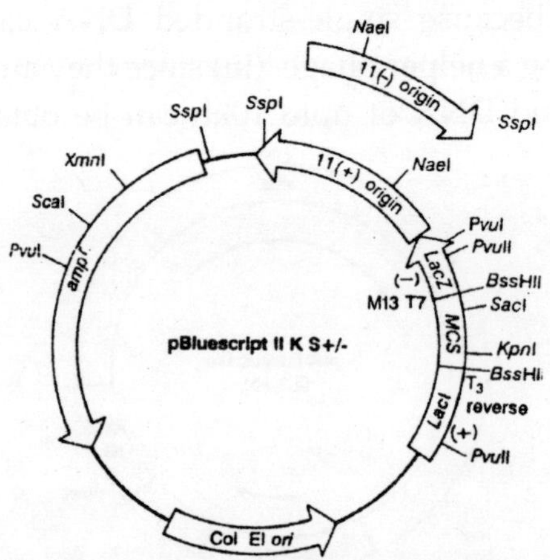

Phagemid pBluescript IIKS +/- and its major characteristics.

that contains promoters from T3 and T7 phages in opposite orientations. The detailed structure of pBluescript II KS (+/-) is given in Figure. It may be noted that it has the following features: (i) a multiple cloning site (MCS) flanked by T_3 and T_2 promoters to be read in opposite directions on the two strands, (ii) an inducible *lac* promoter (/acl), upstream of *lacZ* region, which complements with *E. coli* (lacZ") and provides the facility for selection of chimeric vector DNA (recombinant vector) using the criterion of white colonies (as against blue colonies obtained if no foreign DNA is inserted); (iii) *fl* (+) and *fl* (-) origins of replication derived from a filamentous phage (Ml 3 phage) for recovery of sense single stranded DNA i.e. (+) and anti-sense (-) strands of *lacZ* gene, when host is coinfected with a helper phage; (iv) an origin of replication (ColEl *ori)* derived from plasmid, and used in the absence of helper phage; (iv) a gene for ampicillin resistance for antibiotic for selection of chimeric phagemid vector.

BACs and PACs as Vectors for Cloning Large DNA Segments (100-300 kb) : Genome analysis and map-based cloning in higher plants and animals required isolation of large pieces of DNA, which can not be cloned in plasmid or phage vectors. A variety of vectors are now available, which are suitable for cloning such large DNA fragments suitable for preparation of high-resolution physical maps of higher eukaryotes (both animals and plants). These vectors, 0.5 to 2 megabase (Mb) in length, particularly include yeast artificial chromosomes, (YACs) capable of accepting fragments of 100-2000 kbp and PI vectors capable of cloning fragments of upto 100 kbp (kilobase pairs). Mammalian artificial chromosome (MAC) vectors have also been designed, which are capable of cloning fragments of more than 1000 kilobase pairs. Plasmids based on F-factors were also designed which could accept fragments upto 120kbp in length (for YAC and MAC vector, see later). Of these vectors, YACs are most commonly used, but following difficulties have been encountered while using

YACs : (i) In YAC libraries, a fraction of clones result due to co-cloning events giving single clones with non-contiguous fragments-these are described as chimeric vectors, (ii) YAC clones also exhibit some degree of instability due to deletions/ re-arrangements in the cloned inserts, (iii) YACs are similar to yeast normal chromosomes in size thus making it difficult to separate them by simple methods.

Bacterial Artificial Chromosomes (BACs) : In order to overcome the above difficulties associated with YAC vector, a bacterial cloning system based on *E. coli* F factor was designed which was capable of cloning fragments of upto 300-350kb. These were described as bacterial artificial chromosomes (BACs) and are 'user friendly' being a bacterial system. BAC vectors are superior to other bacterial systems, based on-high-to medium copy number of replicons, since they show structural instability of inserts, deleting or rearranging portions of cloned

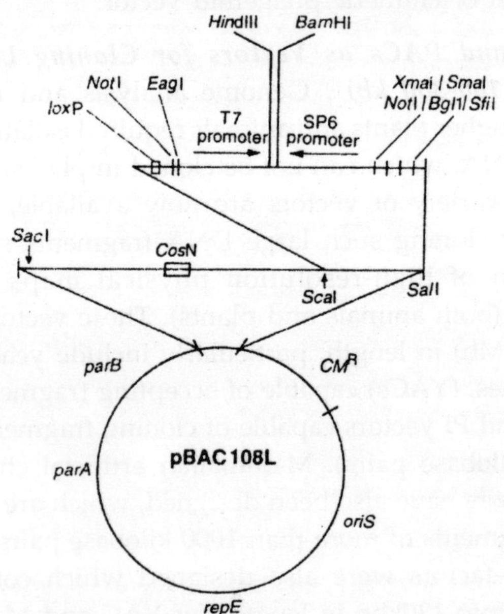

A BAC (Bacterial Artificial Chromosome) vector showing its essential elements for cloning.

DNA. However, the F factor has regulatory genes that regulate its own replication and controls its copy number.

These regulatory genes include (i) *oriS* and *repE* which mediate unidirectional replication and (ii) *park* and *parB*, which maintain the copy number to 1 or 2 per *E. coli* genome. These essential genes of F factor are incorporated in every BAC vector (pBAG), which also has a chloroamphenicol resistance gene as a marker and a cloning segment (Fig.).

Cloning segment includes the following sequences :(i) phage lambda *cosN* site (providing a fixed position for specific cleavage with lambda terminase) and *loxP* site (providing a position for cleavage due to PI Cre protein in presence of *loxP* oligonucleotide); these two sites allow generation of ends that can be used for restriction site mapping to arrange the clones in an ordered array; (ii) two cloning sites *(Hindlll* and *BamHl)* and (iii) several C + G rich restriction sites *(Notl, Eagl, Xmal, Smal, Bgll* and *Sfil)* for potential excision of inserts; the cloning site is flanked by T7 and SP6 promoters for generating RNA probes for chromosome walking and for DNA sequencing. BAC libraries have already been prepared in humans, mouse, rice, wheat, *Lotus*, etc.

P1-derived Artificial Chromosomes (PACs) : These vectors incorporated features of both P1 and F-factor systems and can be transformed into *E. coli* host by electroporation. The PAC vector can accept inserts, 100-300 kb in size. Therefore, in their capability, they are placed between YACs and cosmids. The vector with insert DNA is amplified in *E. coli* and several micrograms of cloned DNA can be recovered from 5-10 ml of the exponential phase of *E. coli* cells. Two P1 cloning vectors with their essential elements are shown in Figure.

Vectors' Role of Plant and Animal Viruses as Vectors : A number of plant and animal viruses have also been used as vectors both for introducing foreign genes into cells and for gene amplification and expression in host cells. In the latter case, we may be interested only in getting increased quantity of the gene product and may not be interested in the integration of the foreign gene in the host genome. Some of the viruses that are commonly used as vectors will be described in this section.

Plant Viruses (Cauliflower Mosaic Virus or CaMV and Geminiviruses) : Cauliflower mosaic virus (CaMV), tobacco mosaic virus (TMV) and geminiviruses are three groups of viruses that have been used as vectors for cloning of DNA segments. CaMV infects particularly the members of Cruciferae and has a double- stranded DNA molecule, 8kbp in size, whose sequence is now known.

Following infection, the virus spreads systematically throughout the plant in a very high copy number reaching upto 105 virus particles per cell. These features make CaMV a suitable vector for transformation of higher plants although there are instances of the use of other viruses also as vectors leading to the production of transgenic plants.

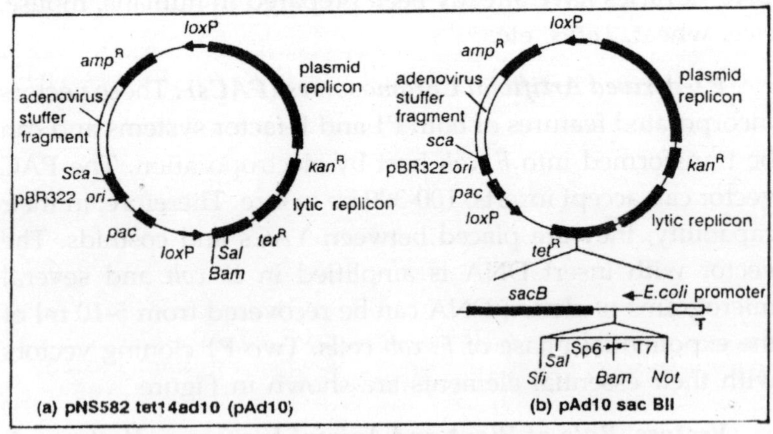

P1 Cloning Vectors with their major characteristics.

Geminiviruses comprise a group of single stranded DNA plant viruses causing important diseases in cassava, maize and other cereals. They replicate via double-stranded DNA forms and have been subdivided into two groups, one infecting monocots and transmitted by leafhoppers and the other infecting dicots and transmitted by whitefly (Table). The dicot geminiviruses have two DNA segments in their genome, DNA A and DNA B, both essential for infection by mechanical inoculation. DNA A carries all genes responsible for replication.

It has been shown that deletions in coat protein gene do not destroy infectivity of the virus, which means that long deletions in this region can be used to insert foreign DNA to obtain gene constructs that will be extremely useful as amplification and expression vectors in plants. Further, since the coat proteins are the most abundant proteins synthesized by the virus, promoters of coat protein genes can be used for high level expression of cloned genes. CLV and TGMV have already been tested for their ability to allow multiplication and expression of cloned genes.

Some important geminiviruses.

Geminiviruses infecting monocots	*Geminiviruses infecting dicots*
1. Maize streak virus (MSV)	1. *Cassia* latent virus (CLV)
2. Wheat dwarf virus (WDV)	2. Tomato golden mosaic virus (TGMV)
3. *Digitaria* streak virus (DSV)	3. Tomato yellow leaf curl virus (TYLCV)
4. *Chloris* striate mosaic virus (CSMV)	4. Bean golden mosaic virus (BGMV)
5. African cassava mosaic virus (ACMV)	

Animal Viruses: A number of animal viruses are also used as vectors, either for the delivery of nucleic acid into cultured cells followed by its integration with host genome or for the amplification and high level expression of foreign genes, using the promoters from virus genes. These cloned genes can have a variety of uses including gene therapy in mammalian cells and synthesis of important proteins by cloned genes in cell cultures. Some of these viruses will be discussed later in this chapter, because they are used as expression vectors.

Role of Vectors Transposons

Transposons of higher plants *(Ac/Ds* or *Mul* of maize). Ac and Ds are popular transposons in corn, and were earlier known to represent activator-dissociation (Ac-Ds) system. Each represents a transposon with short terminal repeats enclosing a long DNA segment, which measures more than 4500 bp in Ac and about 400 bp in Ds. Each possesses genes including the gene for transposase enzyme responsible for transposition. Part of this region can be deleted and the transposon can be used for cloning of foreign DNA segment in much the same way as in other cases.

Transposons of Drosophila (P elements) : P element of *Drosophila* consists mainly of 31 bp (base pairs) terminal inverted repeats enclosing a 3 kilobase protein coding region, coding for transposase and a repressor of transposition. Many P elements carrying internal deletions are also found, where the transposase derived from other complete P elements help their mobilization. In these deleted P elements, foreign gene can be inserted to give recombinant P element, which can be microinjected into the fertilized egg, along with normal P element (helper P to supply transposase). The inserted gene is transposed onto an embryonic chromosome.

YAC and MAC Vectors for Cloning Very Large DNA Segments (1000 kb) : Yeast artificial chromosomes (YACs). We know that in plasmid vectors, sequences upto 10-15 kbp, in lambda (λ) phage vectors sequences upto 22kbp and in cosmid vectors' sequences upto 40kbp can be cloned. Yeast artificial chromosome (YAC) vectors, mentioned earlier allow cloning of sequences that are several hundred kilobase pairs (upto 1000 kbp = 1 Mbp) long (Fig.).

These long molecules of DNA, which may represent whole chromosomes in lower eukaryotes with very small genomes can be cloned in yeast by ligating them to vector sequences that allow their propagation as linear artificial chromosomes. These long DNA molecules can be generated and allow construction of comprehensive libraries (with large DNA segments) in microbial hosts. With the isolation of mammalian telomere and centromere, mammalian artificial chromosomes (MACs) have also been produced.

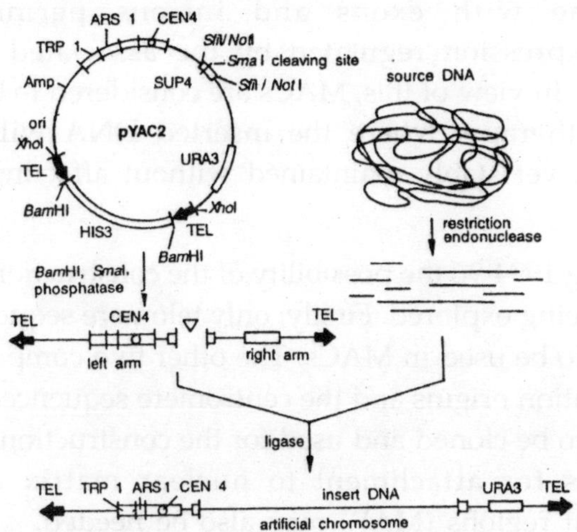

An artificial yeast chromosome vector, which may carry a long DNA segment to be cloned in yeast cells

YACs have two disadvantages, (i) Cloning efficiency is low (1000 clones/fig DNA as against 10^6 -10^7 clones/(Ag DNA for cosmids), thus making it impractical to generate complete genomic libraries through the use of YACs. (ii) It is not possible to recover large amount of pure insert DNA from individual clones; selective amplification of YACs DNA has recently allowed this problem to be overcome. Both these problems have been overcome in BACs and PACs described earlier.

Artificial Chromosomes of Mammals : (MACs) Mammalian artificial chromosome (MAO vectors are designed to be able to replicate, segregate and express in a mammalian cell like any other mammalian chromosome along with other chromosomnes. Since it will be an independent chromosome, with all the functional elements (telomeres, origins of replication, centromere, etc.).

MAC will not be integrated with the genome and can be used as a vector maintaining a single copy per cell. It could carry large fragments of DNA representing an intact eukaryotic split gene with exons and introns permitting its normal expression regulated by the associated promoter sequences. In view of this, MACs are considered to be suitable for gene therapy, where the inserted DNA will be fully expressed, yet stably maintained without affecting the host genome.

During 1994-96 the possibility of the construction of MACs was still being explored. Firstly, only telomere sequences were available to be used in MACs. The other two components, i.e. the replication origins and the centromere sequences were not available to be cloned and used for the construction of MACs. Sequences for attachment to nuclear matrix or matrix attachment regions (MAR) may also be needed.

The elements of YACs are constructed in *E. coli,* ligated *in vitro* and then transferred to yeast cells, but for MACs this may be difficult, since cloning of large mammalian DNA segments in *E. coli* and yeast may be difficult. Therefore, one may design a MAC of the smallest possible size, then construct it in yeast and transfer it to mammalian chromosome. These problems associated with the construction of MACs and their insertion into mammalian cells were partly overcome later. In April 1997 issue of Nature Genetics, successful production of human artificial chromosome was reported. These human artificial chromosomes are 1/5th to 1/10th the size of a normal human chromosome and are already being used for the study of regulation of gene activity and also for gene therapy.

Role of Expression Vectors

In a number of cloning vectors described earlier in this chapter, the main utility has been to clone a gene or a DNA sequence. In other words, they are used for obtaining millions of copies of the cloned DNA segment, for further use in either the genetic engineering experiments or for further basic studies. With this objective in mind, the cloned genes in these vectors need not express themselves either at the transcription level (production of mRNA) or at the translation level (production of protein).

But, when the cloned gene is used for transformation to generate transgenic plants or animals or for production of microbes to be used in industry, then the cloned gene must be expressed. Sometimes, a high level of expression of gene is desirable, if the product of cloned gene is to be recovered as a commercial product. This objective can be achieved through the use of promoters and expression cassettes (an expression cassette is a combination of DNA sequences, which allow faithful expression of the cloned gene).

Promoting Elements

Promoters from non-plant sources express poorly in plants (except promoters of T-DNA) and promoters of non-animal sources express poorly in animals. Therefore, to obtain correct expression, genes transferred to plants should be linked to plant-specific promoters and those transferred to animals should be linked to animal-specific promoters. These promoters can also be interchanged so as to confer on the transferred gene, a specific pattern of expression (tissue-specific or stimulus-specific or development stage-specific expression). To illustrate this, some examples of promoters used in plants and insects will be discussed.

Nopaline synthase *(nos)* promoter from T-DNA. This promoter is 200 bp long and contains several DNA sequence motifs which direct the expression of linked gene. The upstream element from -97 to -130, when duplicated, increases the expression three fold, suggesting that this may be an enhancer. The *nos* promoter is most active in basal regions of plants, its activity slowing down in vegetative parts at the onset of flowering, even though in the flower its activity is increased.

Dual promoter of mannopine synthase *(mas)* genes 1 and 2. These promoters are present in a T-DNA fragment 467 bp long. The expression seems to be developmentally regulated and the linked genes seem to be most active in the basal region of the plant. Expression is also induced on

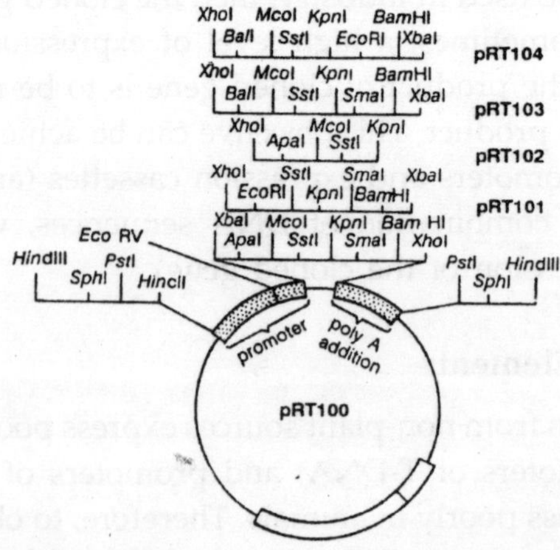

(a) pRT series of expression cassettes based on pUC 18/19

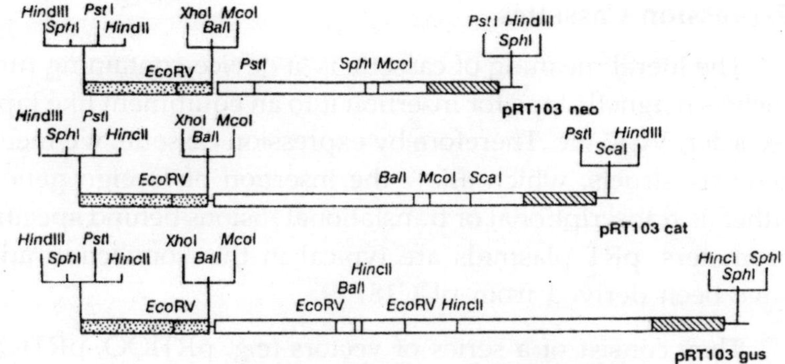

(b) genetic marker cassettes based on pRT103

Expression cassettes for plant cells : (a) pRT series addition sequences from 35S RNA of CaMV : (b) Cassettes with different polylinkers flanked by promoter and poly A with three genetic markers (nptII, cat and gus based on pRT 103.

wounding. Since they are closely linked, if a marker gene is linked to one of these two promoters and a foreign gene is linked to the other promoter, selection for high expression of marker gene will lead to high level of expression of the foreign gene.

35S Promoter of CaMV : This is the most extensively used promoter in a wide variety of plants. The promoter is 343 bp long and contains a strong transcriptional enhancer, which on duplication results in 10-fold increase in expression, even at a distance of 2 kbp. High expression is mainly observed in leaf tissue.

Polyhedrin Promoter from Baculovirus : Baculovirus can synthesize polyhedrin protein-infected cells (baculovirus is an insect virus), so that polyhedrin makes upto 50% of the total protein in the cell, giving upto lg of protein per liter of insect cell culture. In view of this, the promoter has been used to construct expression vectors to allow high level of expression of any gene under the influence of this promoter. This is being used for developing biopesticides or even for production of specific chemicals in industry (see later in this chapter).

Expression Cassettes

The literal meaning of cassette is 'a device-containing film such as magnetic tape for insertion into an equipment like tape recorder, VCR, etc. Therefore by expression cassette, we mean gene constructs, which allow the insertion of foreign genes, either as transcriptional or translational fusions behind specific promoters. pRT plasmids are typical in this connection and have been derived from pUC18/19.

They consist of a series of vectors (e.g. pRTlOO, pRTlOl, pRT102, pRT103, pRT104), which differ in poly-linker sequences, each flanked by 35 S promoter of CaMV on one end and a sequence for poly A addition on the other end (Fig.). A variety of marker genes *(cat, nptll, gus, etc.)* have been inserted into these cassettes and their expression studied, both in protoplasts (transient expression) and in stable transgenic tissues.

Baculovirus and Expression Vector System for Insect Cells: Development of baculovirus expression vector system is based on its life cycle which involves the following steps :

(i) in early phase (12 hours), virus particles are budded from infected cells and spread the virus;

(ii) in the late phase (2-5 days), occluded virus units accumulate in the host nucleus and get embedded in protein polyhedra;

(iii) insect is decomposed releasing polyhedra;

(iv) polyhedra are ingested by another insect host, where polyhedra are dissolved releasing virus for multiplication.

The use of baculovirus as a cloning and expression vector involves following steps (Fig.) : (i) desired gene is cloned in a transfer vector under the control of polyhedrin promoter; (ii) chimeric transfer vector is cotransfected with wild type virus into the host cell; (iii) polyhedrin gene of wild type is replaced by cloned gene through recombination; (iv) recombinant virus is plaque purified, since they will have no polyhedra occlusion bodies; (v) the selected recombinant virus is multiplied and used for expression of heterologus protein.

Virus Expression Vectors for Mammalian Cells : A number of viruses have also been used as transfer or expression vectors for mammalian cells. These include the following : (i) The first viruses used for mammalian cells included DNA viruses like *papovaviruses* (SV40 and polyoma) and *adenoviruses,* but since their capacity for foreign genes is small, they cannot be used for the expression of many genes which are fairly large in size; (ii) *retroviruses* (murine and avian retroviruses) are very useful vectors, since they can infect broad variety of cell types and can be used for cloning large genes; however, being single-stranded RNA (Table), they can not replicate without being integrated so that we will need replicating cells for expression or alternatively high-titer (high concentration) for fully differentiated cells (e.g. neurons, hepatocytes, etc.); (iii) *adeno-associated virus* (AAV) overcomes the limitations associated with retroviruses even through, like retroviruses, it is an integrating virus, with 5kb single stranded DNA; it is

Transforming viruses used as vectors in mammalian cells.

	Viral class	*Type of virus*	*Genome size*
1.	Adeno virus	DNA duplex	37 kb
2.	Papova virus	DNA duplex	5-6 kb
3.	Herpes virus	DNA duplex	160 kb
4.	Adeno associated virus	DNA (single strand)	5 kb
5.	Retro virus strand)	RNA (single	6-9 kb

Agrobacterium-based plant transformation vectors.

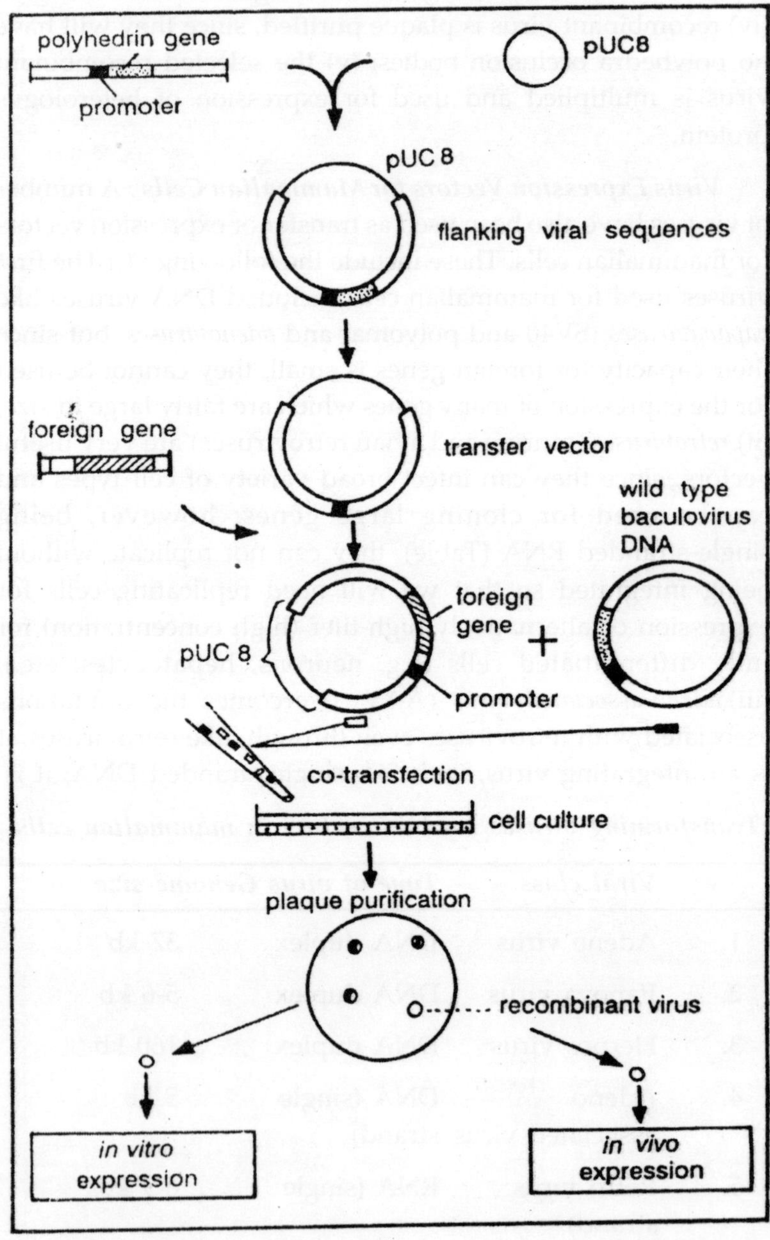

Construction of baculovirus expression vectors.

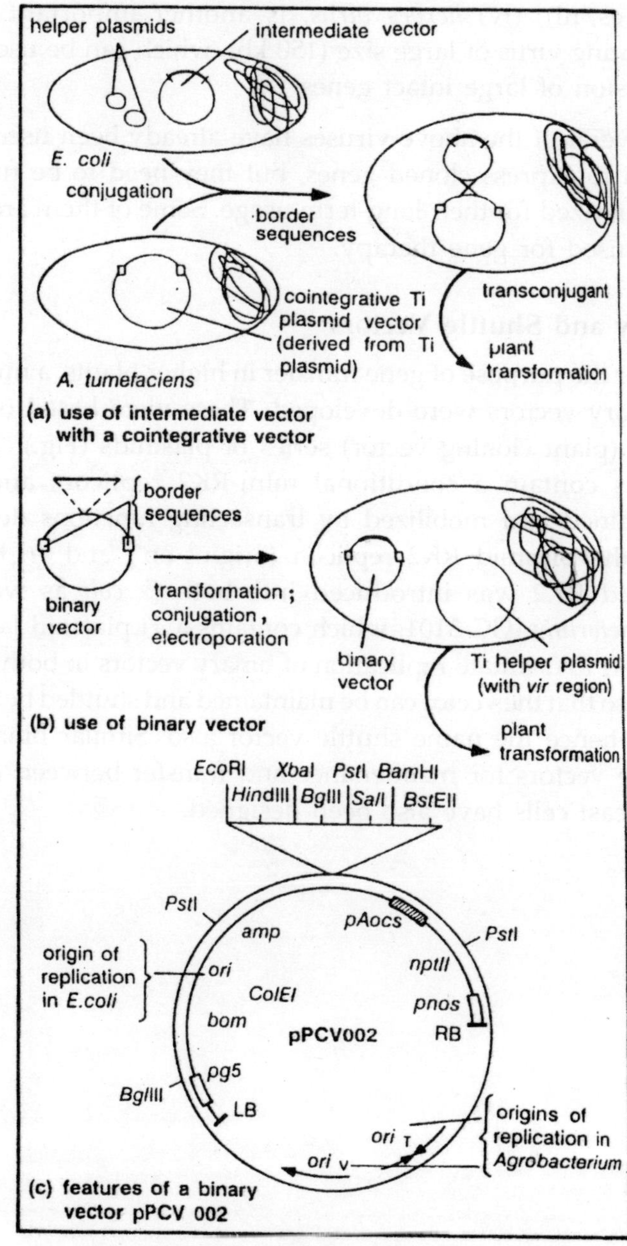

Agrobacterium based plant transformation vectors.

non-pathogenic, and can be concentrated to high titer (109 particles/nl); (iv) *herpes virus,* is another important non-integrating virus of large size (150 kb), which can be used for expression of large intact genes.

Several of the above viruses have already been used and shown to express cloned genes, but they need to be further characterized for their long-term usage. Some of them are also being used for gene therapy.

Binary and Shuttle Vectors

For the purpose of gene transfer in higher plants, a number of binary vectors were developed. These were based on the pPCV (plant cloning vector) series of plasmids (Fig.). These vectors contain a conditional mini-RK2 replicon, and are maintained and mobilized by transacting functions derived from the plasmid RK2 replicon (origins-on'$_v$ and *ori$_T$*). The plasmid RK2 was introduced into both *E. coli* as well as *Agrobacterium* (GC 3101, which contains a Ti plasmid lacking T DNA) to facilitate replication of binary vectors in both these hosts, so that the vector can be maintained and shuttled between both—hence the name shuttle vector also. Similar binary or shuttle vectors for maintenance and transfer between *E. coli* and yeast cells have also been designed.

7

Variety of Mechanisms

During the last two decades, considerable information has been made available on the regulation of gene expression in eukaryotes, although in the 1960's only bacteria were considered suitable for such studies. Due to inherent difficulties in working with multi-cellular eukaryotes, initial studies in these systems were restricted to the following : (i) formulation of theoretical models for the regulation of gene expression and (ii) conducting experiments to identify whether histones or non-histone chromosomal proteins are relatively more important for the regulation of the activity of specific genes in time and space (i.e. at specific developmental stage and in specific tissue). More recently (particularly after 1980), regulation of specific genes has been examined in greater detail giving valuable information.

Specific DNA sequences and proteins (e.g. transcription factors) involved in regulation of specific genes have been identified. These specific genes express either in specialized cells or respond to specific stimuli and are variously described as luxury genes or smart genes as opposed to house keeping genes, which are constitutive in their expression. A large number of transcription factors have been identified which take part in the formation of a transcription complex, which activates the enzyme, RNA polymerase and associated proteins to initiate transcription. Specific regulatory DNA sequences have also been identified (sometimes through the use of transgenic plants), which respond to stimuli and provide sites for binding of transcription factors. For transmission of stimuli (external or internal) to the target DNA sites, specific signal transduction pathways (involving intermediate receptor molecules) have been worked out in detail in many cases. Examples are also available now, where regulation of gene expression in eukaryotes is exercised at the post-transcriptional level involving RNA processing or translation. A remarkable phenomenon of regulation through genome reorganization or gene rearrangement has also been discovered in (i) human immune system involving antibody production, (ii) yeast mating types involving production of pheromones-receptors and in (iii) trypanosome's variable surface antigen-glycoprotein (VSG) production.

Another important area of research, which is related with regulation of gene expression in eukaryotes is cell signalling. The cells do possess structures and mechanisms, which permit them to perceive stimuli and transmit the signal to the interior of the cell. Consequently, appropriate genes are expressed at appropriate time and in appropriate tissue. The cells possess receptor molecules on their surface or inside the cell, which help in receiving the signals, and also possess signal transduction pathways which transduce the signals to the genes, whose expression needs to be regulated. This area of research has been actively pursued during 1980's and 1990's and spectacular success has been achieved. All these aspects will be briefly described in this chapter. Limited space in this book does not allow detailed treatment of the subject, which has expanded rapidly in recent years.

Transcription-level Regulation

As in prokaryotes, regulation of gene expression is exercised in majority of cases at the level of transcription, although examples of post-transcriptional control are also known. In view of this, major part of this chapter is devoted to regulation at transcriptional level. This regulation may involve both activation and repression, because genes may have to be switched on in some cases and switched off in other cases. DNA sequences involved in regulation were also identified through the use of DNA footprinting as well as through the use of gene constructs for production of transgenic plants.

Transcription Activation : Activation of transcription as a means of regulation of gene expression has been studied during the last more than 20 years, using techniques, which have undergone a revolution through the development of recombinant DNA technology during 1980's In a chronological order, initially (1969-73), these studies involved formulation of theoretical models based on the occurrence of large proportion of repetitive DNA in eukaryotes. Later (1973-75), chromatin reconstitution experiments demonstrated a positive control exercised by non-histone chromosomal proteins and still later (particularly in 1980's and 1990's), the role of transcription factors was examined in detail in large number of cases.

Britten-Davidson's Model for Unit of Transcription : Although various models for regulation of protein synthesis in eukaryotes have been proposed, a popular model, known as Britten-Davidson model or the gene-battery model was one of the first models which was proposed by two scientists, Britten and Davidson in 1969. It was subsequently elaborated by them in 1973. The model assumes the presence of four classes of sequences, (i) Producer genes. A producer gene is comparable to a structural gene of prokaryotic operon. (ii) Receptor site. A receptor site is comparable to operator gene in bacterial operon and one such receptor site is assumed to be present adjacent to each producer gene or a set of such producer genes, (iii) Integrator gene. It is comparable to regulator gene and is responsible for synthesis of an activator RNA that may or may not give rise to proteins before it activates the receptor site (Fig.).(iv) Sensor site. A sensor site regulates the activity of integrator gene, which can be transcribed only when the sensor site is activated. The sensor sites are recognized by agents which like hormones and proteins change the pattern of gene expression. For instance, hormone-protein complex or a transcription factor may bind to a sensor site and cause the transcription of integrator.

The inter-relationship of the four classes of sequences given above are illustrated in Figure. In this model, the genes (producer gene and integrator gene) are those sequences which are involved in RNA synthesis. On the other hand, sites (receptor and sensor sites) are those sequences which help only in recognition without taking part in RNA synthesis. It is also proposed that receptor sites and integrator genes may be repeated a number of times, so as to control the activity of

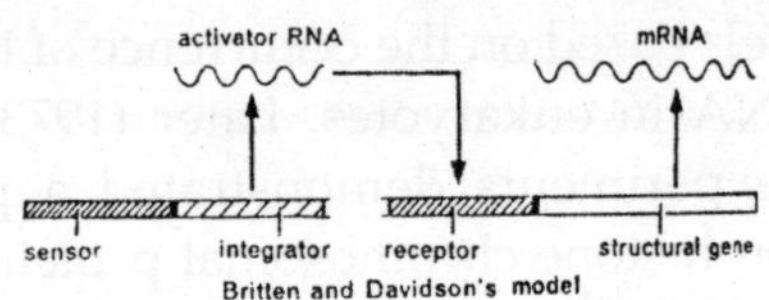

Different components in Britten-Davidson's model.

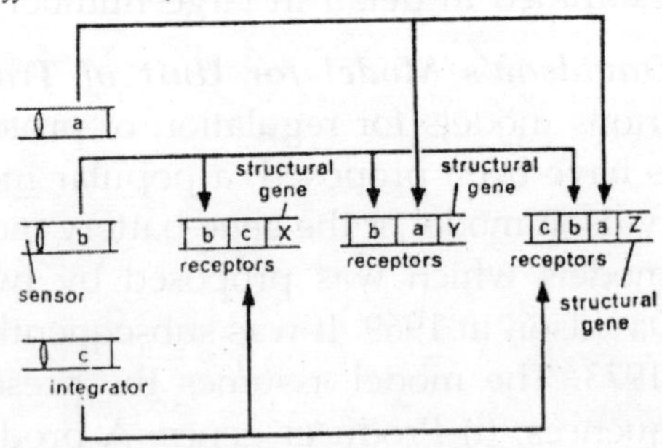

Redundancy of receptor sites (as proposed in Britten-Davidson's model)

a large number of genes in the same cell. Repetition of receptor ensures that same activator recognizes all of them and several enzymes of one pathway are simultaneously synthesized.

It is also possible that the transcription of same gene be needed in different developmental stages. This can be achieved by multiplicity of receptor sites and integrator genes. Each producer gene may have several receptor sites, each responding to one activator so that a single activator can recognize several genes, but different activators may activate the same gene at different times. An integrator gene may also fall in cluster with same sensor site.

Gene Battery : In Britten-Devidson model, a set of structural genes controlled by one sensor site is termed a battery. Sometimes, when major changes are needed, it is necessary to activate several sets of genes. If one sensor site is associated with several integrators, it may cause transcription of all integrators at the same time thus causing transcription of several producer genes through receptor sites. The repetition of integrator and receptor sequences is consistent with the knowledge that sufficient repeated DNA occurs in eukaryotes.

Chromosomal Proteins and Gene Expression : As we know, chromosomal proteins are mainly of two kinds : (i) histones and (ii) non-histones, so that the chromatin mainly has three components— DNA, histones and non-histones. While it is known since early 1960's that histones may be involved in repressing gene activity, the specific regulation by non-histones was demonstrated only during 1970's. The most important

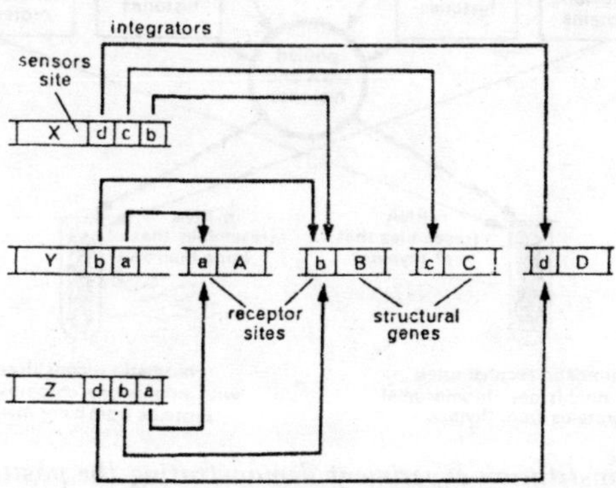

Redundancy of integrator genes (as proposed in Britten-Davidson's model)

experiments in this connection were pioneered by Gilmour and Paul (1973). They isolated the chromatin from different tissues separately and then dissociated it into DNA, histones and non-histones. This was followed by chromatin reconstitution using either the three components derived from chromatin of the same tissue or by combining non-histones of one tissue with the DNA and histones of another tissue (Fig.). It could be demonstrated in such experiments that the mRNA which is synthesized in vitro from reconstituted chromatin, mainly depended on the source of non-histone proteins (for a review consult O'Malley et al., 1977).

There are at least three specific proteins, whose synthesis in eukaryotes has been studied using chromatin reconstitution experiments. These are (i) the globin synthesized in reticulocytes, (ii) the ovalhumin synthesized in chick oviduct due to hormonal stimulation and (iii) the histones synthesized

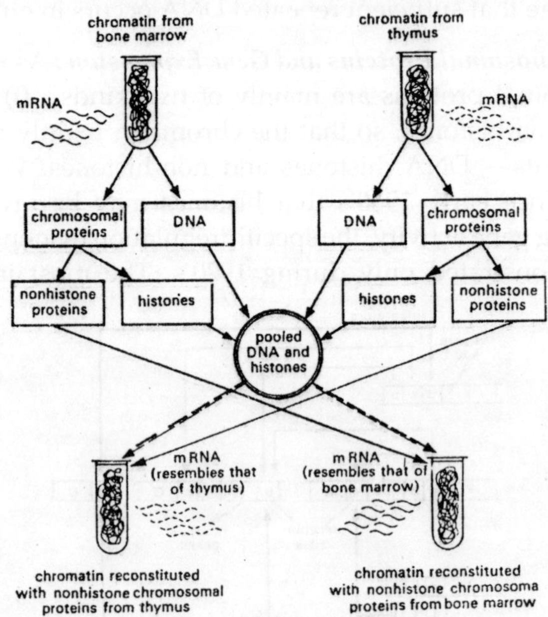

Chromatin reconstitution experiment demonstrating the positive role of non-histone proteins.

during 'S' phase of cell cycle. In all these three cases, similar observations were available which are as follows :

(i) Same quantity and quality of mRNA was synthesized on native chromatin as well as on chromatin reconstituted from DNA, histones and non-histones derived from same tissue.

(ii) When heterologous chromatin was reconstituted by using non-histones from one tissue and DNA and histones from another tissue, mRNA synthesized was similar to that synthesized on the native chromatin to which non-histones belonged.

Above observations suggested a positive regulatory role of non-histones in eukaryotes. This role of non-histones is also suggested by the following additional information,

(i) Non-histones are present in increased quantity in tissues active in RNA synthesis, while quantity of histones is constant,

(ii) Non-histones exhibit 2-3 times more diversity than histones,

(iii) Non-histones have tissue specificity and DNA-binding specificity,

(iv) Certain non-histones are known to induce RNA synthesis in vitro,

(v) Synthesis of specific non-histones is associated with induction of gene activity,

(vi) A large number of transcription factors.

Above observations though suggest a specific role of non-histones, but in the absence of histones, non-histones have been shown not to induce RNA synthesis. This suggests that regulation of protein synthesis actually involves an interaction between histones and non-histones, where histones inhibit protein synthesis non-specifically and non-histones induce RNA synthesis rather specifically. Some histone proteins may also function as transcription factors.

Transcription Complex and Activating Smart Genes : It has now been established through a variety of studies, that for the elaborate transcription process to get underway, several proteins (called transcription factors) must bind to the DNA sequences near a target gene. This forms a transcription complex, which works as 'brain' of a smart gene and is comparable to a 'computer'. In this brain or computer, the signals are combined to make a decision about whether or not to switch on the smart gene. Only by the right combination of transcription factors or signals, the transcriptional enzyme (RNA polymerase) is activated, and produces RNA. In contrast to this computer-like complex control of transcription in eukaryotes, the control in prokaryotes is simple and can be compared to mere switches.

In eukaryotes, a number of DNA-binding proteins and their corresponding binding regions on DNA have been identified and isolated in a variety of organisms. While in prokaryotes, there is a maximum of three protein-binding sites near the transcription start point, in eukaryotes, these sites may be at least five and may be as many as 20 for some smart genes (not for house keeping genes). For instance, in sea urchin, for the regulation of a gene producing actin, as many as 20 DNA regions near this gene have been identified, which are recognized by regulatory proteins. At least five of these 20 sites are used for activation of actin gene.

We described seven transcription factors (TFIID, TFIIA, TFIIB, TFIIE, TFIIF, TFIIJ, TFIIH), which are essential for transcription of all genes including house keeping genes and smart genes. However, for transcription of smart genes, additional transcription factors must form a complex to activate the obligatory proteins coupled to the RNA polymerase II enzyme that transcribes the gene into hnRNA. Only when all components (including the obligatory proteins and the right combination of transcription factors forming a complex) are present, the transcription of a smart gene will take pace. Several hundred different transcription factors have been found to date, two-third of these falling into four recognizable groups—helix-turn-helix, helix-loop-helix, zinc finger and leucine zipper. The remaining one-third may be unique to their types. Each transcription factor has separate domains for (i) recognition and binding a specific DNA sequence and for (ii) activation of obligatory proteins. These are called DNA-binding domain and activation domain.

Some of the earliest studied transcription factors include GAL4 and GCN4 in yeast, which activate genes involved in metabolism of galactose and those involved in general control of nitrogen metabolism (amino acid biosynthesis) respectively. The details of these two transcription factors are shown in Table.

GAL4 and GCN4 showing DNA binding and activation domains.

Protein	*Total length*	*Domains*	
		DNA binding	*Activation*
GAL4	881	1-74	147-196 & 768-881
GCN4	281	1-60	107-125

A number of other transcription factors are listed in Table. The mechanism involved in binding of these transcription factors to DNA and their involvement in protein-protein interactions. It should be recognized that the discovery of regulatory DNA sequences and transcription factors was done independent of each other in many cases, so that in all cases it is not known, which transcription factor regulates the activity of which specific genes. However, certain transcription factors are known to be found in some specialized cells and not in others, so that some idea about the genes regulated by them can be inferred. In other cases, where genes have been isolated, their regulatory sequences could also be identified. Similarities are available among the different regulatory DNA sequences as well as among amino acid sequences of different transcription factors, suggesting mechanisms of DNA-protein interactions.

Some transcription factors and then properties.

Transcription factor	*DNA binding domain*	*Specific function*
GR	Zn finger	Glucocorticoid receptor
Sp 1	Zn finger	General
CTF/NF-1	—	General
C-Jun	Leucine zipper (for dimerization)	Binds AP-1 sites
Fos	Leucine zipper (for dimerization)	—
C/EBP	Leucine zipper (for dimerization)	Hepatocytes and fat cells
AP2	—	Confers TPA and cAMP inducibility
CREB	Leucine zipper	Confers cAMP inducibility and El A transactivation
OCT-1	Homeodomain	Binds octamer; general
OCT-2	Homeodomain	Binds octamer; lymphoid
SRF	—	Serum
NF-κB	—	Lymphoid
HSTF	—	Heat shock

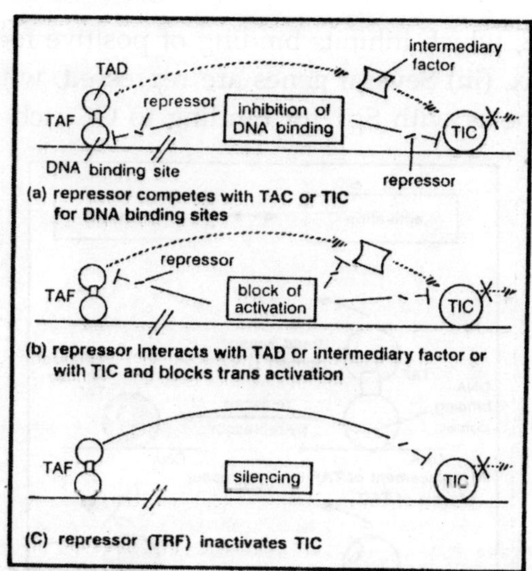

Three categories of transcriptional repression (TAF = trans-activation factor; TIC = transcription initiation complex; TRF = trans-repressing factor).

Repression of Transcription : In recent years, many examples of negative control of transcription have also become available in eukaryotes. These examples can be divided into three main classes:

(i) inhibition of DNA binding,

(ii) blocking of activation and

(iii) silencing. These three mechanisms of repression are illustrated in Figure.

Inhibition of the Binding of Trans-activating Factors (TAF) to DNA : One of the first examples of inhibition of TAF binding involved SV40 T antigen binding to its own promoter as in prokaryotes. In many other cases, there is a competition between a negative factor and a positive TAF as shown in the following examples : (i) In sea urchin, sperm-specific histone gene H2B-1 is repressed in embryos, due to interference by a negative factor with a positive CCAAT binding factor, (ii) In humans, fetal y-globin gene is repressed in adults due to binding of a factor NF-E, which inhibits binding of positive factor CP1 to CCAAT box. (iii) Several genes are repressed, when GC-box factor competes with Sp1 for binding to GC-rich sequences.

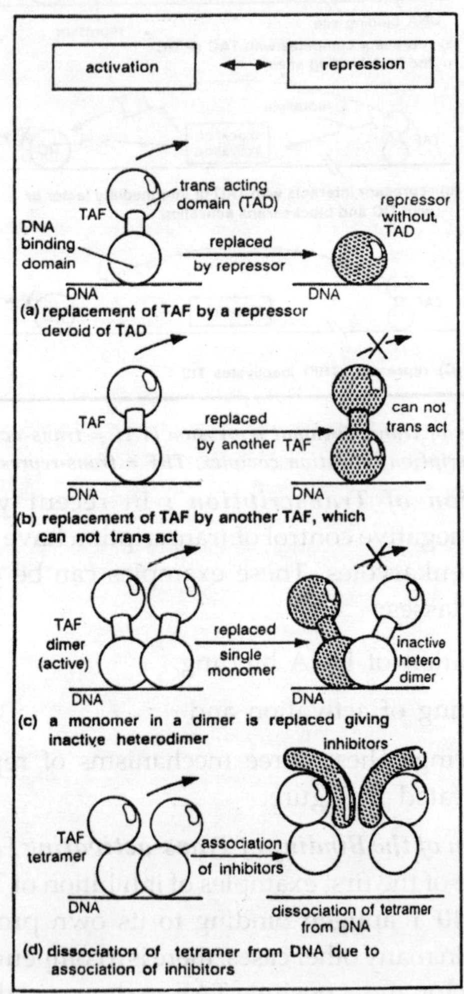

Four ways in which DNA binding of TAF can be inhibited.

During such an inhibition, a positive factor for one gene may also function as negative factor for another gene due to similar DNA sequences. This has been shown in several developmental genes in Drosophila and also for glucocorticoid receptor molecule, (iv) In still other cases, a component (e.g. Fos) of a heterodimer (e.g. Jun-Fos) can be replaced to give inactive complex (e.g. C-Jun-Jun B), which competes with heterodimers; or an inhibitor (e.g. IkB) may cause a TAF dissociate from DNA and inactivate it (Fig.). Heat shock protein (hsp90) also associates with glucocorticoid receptor and renders it incapable of binding DNA.

Activation Blocking : In several examples, TAF can bind DNA, but its activation domain is not allowed to function (Fig.). Following alternatives are possible :

(i) Negative factor may bind near the binding sites meant for positive factors. For instance, in c-myc gene, the negative factor myc-PRF binds next to positive factor myc-CFl, and for polyoma virus, repressor PEA2 binds near PEAl.

(ii) Negative factor may bind the positive factor, rendering its trans-activation domain ineffective as done by binding of GAL-80 to TAF GAL4 in yeast; similar is the situation with binding of AP-1 with P67SRF needed for activation of C-fos gene (AP-1 is usually a TAF, but here it works as a repressor).

(iii) In signal transduction, sometimes an intermediary protein is involved, which is blocked by another trans-activating domain, which may be present in high concentration (e.g. receptors of several steroid hormones). El a (an adenovirus protein immortalizing primary cells) also blocks activation of several genes by binding intermediary proteins involved in activation due to TAFs (AP-1, GAL4, VP16), which bind to SV40 enhancers.

Silencing of Transcription : After the discovery of enhancer elements involved in enhancing transcription in 1985, elements were discovered in yeast, which mediated repression and therefore were called silencers. Like an enhancer, a silencer also functions irrespective of its position (upto 2.5kb away) and orientation relative to the gene, whose expression it controls. The role of silencers can be illustrated using the example of mating type loci in yeast, where a mating type active locus MAT is associated with two silent loci, one HMR on the right side and the other HML on the left side. These silent loci can be 'α' mating type or 'α' mating type and can replace MAT

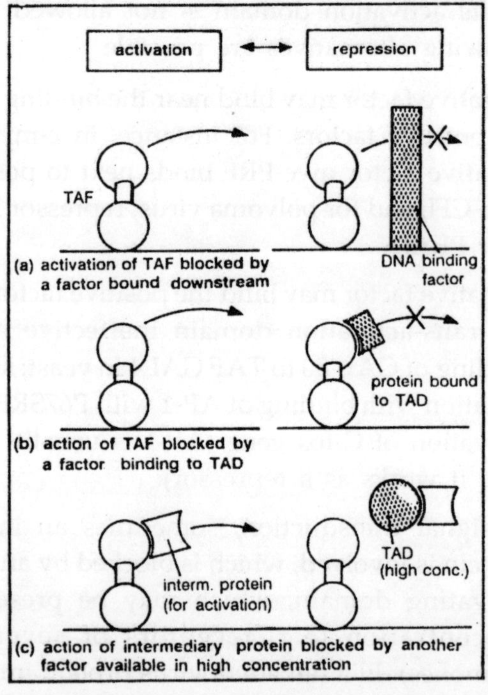

Three ways in which activation (not binding on DNA) by TAF is blocked.

to bring about a change in mating type (see later in this chapter). The loci HMR and HML are each repressed by a silencer located ~ 1 kb upstream in each case and described as E and I silencers or as ER (or HMR-E) and EL (or HML-E). Besides, E and I silencers, there are several transacting factors involved in silencing. For instance, HMR-E has been shown to have binding sites for a transcription factor RAP-1 and another protein Abf-1p needed at replication origin. Both proteins are needed for repression. Similarly α_2 protein coded by one of the genes at HML locus binds to *HML-E* and represses mating type gene(s). There are also four SIR genes (SIR1, SIR2, SIR3, SIR4) involved in silencing. Each of these four genes have been cloned and their products are believed to take part in silencing by interacting with histones to modify chromatin structure.

Silencer sequences have also been reported for human immunodeficiency virus (HIV) and called negative response element (NRE), and also for chicken lysozyme gene. The silencer factor (a protein) either locks the transcription initiation complex and makes it unavailable for activating factors or it disorganizes the transcription initiation complex (Fig.).

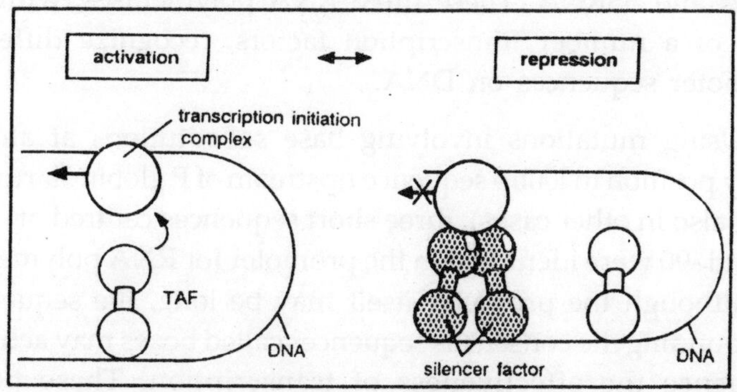

Mechanism of the action of silencer factors, rendering the transcription initiation complex ineffective.

It should be realized that repression of gene activity is not mediated by a single mechanism and one or more of the above three mechanisms may be involved in the same case or even additional mechanisms may be involved. It has been shown that the steroid receptor superfamily utilizes all the three mechanisms for repression and is also involved in activation of transcription although of different genes. More information in this connection will made available in coming years.

Specific-DNA Sequences Regulating Transcription : A large number of promoter DNA sequences have been identified with a variety of genes in eukaryotic genomes. These regulatory DNA sequences may be present either upstream (before the start site) or downstream (after the start site), and used for binding of a number of proteins, most of them being transcription factors. Some of these sequences are present in all genes, but there are others, which are specific and are therefore used for regulation of gene expression.

Upstream DNA sequences controlling transcription by RNA polymerase II in eukaryotes (TATA box, CAAT box and GC box). Eukaryotes have three RNA polymerases (I, II, III). RNA polymerase synthesizes rRNA, RNA polymerase II transcribes protein coding genes and RNA polymerase III synthesizes tRNA and 5SRNA. These three RNA polymerases, with the help of a number transcription factors, recognize different promoter sequences on DNA.

Using mutations involving base substitutions at almost every position in 100bp sequence upstream of P globin startpoint (and also in other cases), three short sequences centred at -30, -75 and -90 were identified in the promoter for RNA polymerase II. Although the promoter itself may be long, the sequences surrounding the consensus sequences called boxes may actually influence the effectiveness of transcription. These boxes themselves are small in size and common in many tissues and organisms. The best characterized of these modules called boxes are TATA box, CAAT box and GC box (Fig.).

TATA box has a resemblance with Pribnow box of prokaryotes and was identified, in the laboratory of Dr. Hogness, by Dr. Goldberg during his Ph.D. work. It is invariably AT rich, Therefore it is often referred to as Goldberg-Hogness box or TATA box. It has been shown that the presence of TATA box is essential for transcription.

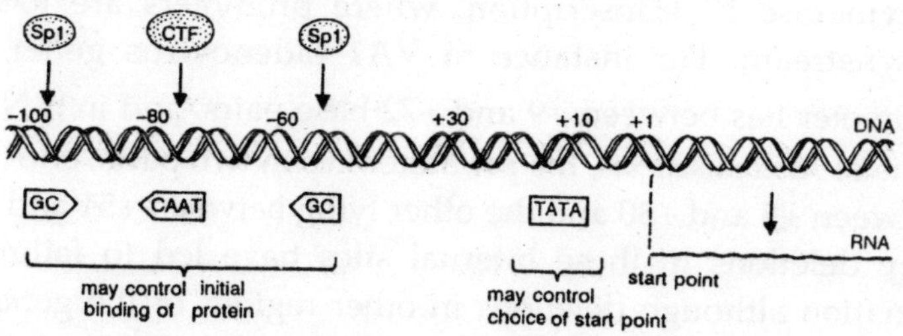

Different consensus sequences (boxes) in the promoter region for thymidine kinase gene (orientation of sequence is shown by arrows; SplandCTF are transcription factors binding to GC and CAAT boxes).

This has been proved with the help of both in vitro (in culture) and in vivo (living system) experiments, where deletions or mutations in TATA box resulted in reduced or complete halt of transcription. However, this is the least effective component of the promoter. It has also been shown that other sequences both downstream from TATA box (between TATA box and initiation site) and upstream from TATA box upto a distance of 200 bases may be important in controlling transcription or in other words, the gene activity.

Further upstream from the transcription initiation site, at position between –70 and –80, is also found a consensus sequence, which has the pattern, GG (C or T) CAATCT. It is called CAAT box, which is the most effective component of the eukaryotic promoters. The GC box, which may occur between–50 and –60 or between –90 and –100 contains the sequence GGGCGG and may be found in multiple copies in either orientation.

Promoter of Downstream for RNA Polymerase III : Promoter sequences for transcription have often been located only upstream the starting point. However, the genes coding for 5SRNA in Xenopus luevis have been shown to have promoters downstream the initiation site and within the transcription unit between +55 and + 80 base pairs from the startpoint of the gene. There are also other cases of RNA polymerase III transcription, where promoters are located downstream. For instance in VA1 (adenovirus gene), the promoter lies between +9 and +72 base pairs, and in $tRNA_f^{met}$ gene of *Xenopus laevis,* the promoter lies in two parts, one lying between +8 and +30 and the other lying between +51 and +72. Any deletions in these internal sites have led to failure of initiation although deletions in other regions of the gene did not cause any failure of initiation. Some other IRNA genes in eukaryotes also have similar promoter regions.

Homeotic Box and Regulation of Gene Activity in Drosophila and other Eukaryotes : During 1984 and 1985, a set of genes called homeotic genes, which determine the body plan, received major attention of molecular biologists. These homeotic genes are such that mutations in these genes cause clean transformation of one body part into another. In this connection, in Drosophila, two gene complexes, namely bithorax and antennapedia (antenna-foot) have been studied in some detail and have now been cloned.

In each of these two complexes, there are several genes, and not all of them are homeotic. It has been demonstrated that in almost all these genes, homeotic or non-homeotic, a nucleotide sequence (180 base pairs long) is always found near the 3' end of the gene. This sequence seems to be highly conserved and has now been discovered in other organisms like Xenopus (toad), mice and humans (in humans, at least five such homeotic genes having this 180 base pair long sequence are known). This sequence (180bp long) has been named Homeo box by Walter Gehring of the University of Basel in Switzerland. Some of these genes will be described.

In Drosophila, a number of genes of antennapedia complex include *'Antp' (antennapedia), 'ftz' ifushi tarazu)* 'engrailed'' and 'deformed'. Similarly genes of bithorax complex include ubx or ultrabithorax and infrabdominal-2. Among these genes '*ftz*' genes are interesting and are characterized as segmentation genes that follow the 'pair rule', since every alternate segment is missing thus reducing the number of segments to half (*ftz* homozygote, which dies after early larval stage; fushi tarazu is a Japanese word meaning not enough segments). In homozygous individuals for mutation in 'engrailed' gene also, the alternate segments are defective thus following 'pair rule'.

Similarly, 'Antp' gene in its mutant form, in homozygous condition, causes the conversion of antenna of adult flies into legs. These genes can prove useful in the study of genetic control of developmental pattern.

In some of the above genes, 'homeo box' has been sequenced as in 'fushi tarazu' (*ftz*). It has been shown that although the genes having homeo boxes may widely differ in their phenotypic effect to the extent that some of them are not really homeotic, but the 'homeo box' sequence seems to be similar and highly conserved.

Therefore, it is believed that their role in regulation of gene activity should be similar. The 60 amino acids coded by 'homeo box' seem to give rise to a domain within the protein products of each of the homeotic genes that constitute the complexes like bithorax and antennapedia. It has also been shown that the 'homeo boxes' of Drosophila genes are homologous to yeast *MAT* α 1 and *MAT* α 2 genes, which regulate sets of genes determining the mating type of yeast cells (see later in this chapter for more details). The protein products of *MAT* α 1 gene of yeast and that of 'engrailed' gene of Drosophila have been shown to bind specifically to a specific' DNA sequence thus regulating the activity of either those genes to which the homeo box belongs or even other unrelated genes.

Factors of a Response for a Variety of Stimuli : In signal transduction pathways (see later), smart genes are expressed eventually due to the presence of unique modules in DNA. These modules represent short consensus sequences, usually < 200 bp upstream of the gene to be expressed. These are called response elements (RE) and may be present (either in the promoter or in the enhancer) in single or multiple copies. The response elements are used by certain proteins (e.g., receptor protein or transcription factor), the binding region of which extends for a short distance on either side of RE. Some examples of response elements representing these proteins are given in Table.

Ubiquitin Protein and Regulation of Heat Shock Genes : A massive increase in the transcription of DNA sequences included in heat shock genes has been observed due to increase in temperature. Therefore, heat shock genes offer an ideal system for the study of regulation of gene activity. These genes occur in wide range of organisms ranging from bacteria to humans. It has been shown that a protein named ubiquitin plays a key role in this regulation.

The promoters of eukaryotic heat shock genes possess a response element called heat shock element (HSE) (Table) located upstream of these genes. The HSE is recognized by a 'heat shock transcription factor (HSTF)', which is active (perhaps through phosphorylation) only during heat shock and binds to a site (70bp) carrying HSE (15bp). About 20 genes carrying same HSE are switched on simultaneously, producing heat shock proteins. Both HSE and HSTF have been conserved during evolution. For instance, in yeast and fruitfly similar HSTF proteins and similar HSE (DNA sequences) have been found. It is suggested that perhaps the heat shock transcription factor (HSTF) itself remains ubiquitinated under normal condition and that failure of ubiquitination renders this factor free to be available to attach to a sequence (HSE) meant to initiate transcription of heat shock genes. The role of uhiquitin in modification of nucleosome structure through its association with histone proteins has also been suggested.

Some Response Elements on DNA, which are Recognized in Response to Specific Stimuli in Eukaryotes.

Agent of stimulus	*Factor & size (daltons)*	*Module & consensus*	*DNA bound*
1. Heat shock	HSTF (93,000)	HSE (CNNGAANNTCCNNG)	27bp
2. Glucocorticoid	Receptor (94,000)	GRE (TGGTACAAAGTTCT)	20bp
3. Cadmium	? (?)	MRE (CGNCCCGGNCNC)	?
4. Phorbolester (TPA)	API (39,000)	TRE (TGAC TCA)	22bp
5. Serum	SRF (52,000)	SRE (CCATATTA-GG)	20pb
6. Light	—	LRE	—

Transgenic Plants to Study Regulatory Sequences : The response elements have often been identified through the use of gene constructs, consisting of sequences flanking the gene in question and a reporter gene (e.g. cat gene for chloramphenicol acetyltransferase or gus gene for β glucuronidase). These gene constructs are used for getting transgenic plants where the response of reporter gene under the control of RE is assessed.

Transgenic plants have particularly been found useful in the study of development, since with their help, gene expression can be studied in time and space i.e. in different environmental conditions causing induction or suppression of gene expression. Transgenic plants thus facilitated the recognition of regulatory sequences involved in these differential expressions of gene activity. For this purpose, sequences upstream to a number of structural genes were cloned either with their own structural genes or with other unrelated structural genes (e.g. reporter genes) to study the specific regulatory role of these sequences in time and space. A number of structural genes whose regulatory sequences were examined in this manner are listed in Table, and the details of some of these are described in the following text.

Gene for Small Subunit of Ribulose Bisphosphate Carboxylase (rbcS) : Dozens of genes in plants are affected by light and many of these regulated by phytochrome. The expression of phytochrome gene is depressed and that of RUBISCO (ribulose, 1,5 bisphosphate carboxylase/oxygenase) gene is stimulated by red light. RUBISCO is involved in CO_2 fixation and is the most abundant enzyme on the earth. It consists of two subunits, each encoded by a separate gene, the smaller subunit rbcS encoded by a nuclear gene and the larger subunit rbcL encoded by a chloroplast gene. Although, rbcS from different plants are fairly similar, small differences, that exist, can be used to distinguish mRNA of a transgene from that of the host plant using specific probes. The pea rbcS gene was transferred to petunia and tobacco plants, and soyabean rbcS gene was transferred to petunia. The gene expressed itself in transgenic plants, rbcS subunit from the introduced gene sometimes even combining with rbcL subunit of the host plant to yield the enzyme. 'Deletion mutants' having sequences deleted upstream of the transgene rbcS led to the identification of a sequence' upstream (–166 to –149 bp) of TATA box, which was light responsive and was, therefore, conveniently called light responsive element (LRE). This LRE sequence could also regulate genes other than rbcS gene, so that the activity of these unrelated foreign genes became light, organ and age specific in a manner similar to rbcS gene (high expression in leaves, low expression in stem and no expression in roots). LRE was shown to have three regions, boxI, boxII and boxIII and it was shown that box II region also had a negative regulatory element. This LRE also seems to be repetitive (with two repeats) and each repeat unit can mediate light responsiveness and organ specificity. However, it has also been shown that both repeats of LRE are required for full expression of the rbcS gene in youngest leaves of tobacco plants.

It has also been shown using transgenic plants, that a different sequence upstream to rbcS is responsible for age responsiveness. For example, it was shown that the region from –410 to –166 bp is essential for high level of expression in immature leaves, but is not important for light responsive expression in mature leaves, where the region –166 to –149 bp was most important.

Gene for Chlorophyll a/b Binding Protein (Cab). Examples of the transfer of Cab gene from pea, wheat and Nicotiana *plumbaginifolia* are listed in Table. In this case it was shown that a 247 bp sequence for pea CabAB80 (–347 to –100) and a 268 bp sequence (–357 to –89) for wheat Cab-1, when separately fused with heterologous TATA sequence fused to neomycin phosphotransferase (NPTII) coding sequences, conferred on this coding sequence a light responsive expression. This regulatory sequence when present induced expression of NPTII by red light and the expression could be reversed by a flash of far-red light as observed in phytochrome mediated expression. In another experiment, negative elements suppressing expression were also identified, when *Cab-E* gene of *Nicotiana plumbaginifolia* was studied in a similar manner in transgenic plants.

Heat shock genes : Thermal stress induces the transcription of a small set of heat shock genes in diverse organisms, and also reduces the expression of many other vital genes. The upstream region of heat shock genes regulating their expression is highly conserved, so that when a gene construct carrying NPTII reporter gene, fused with upstream region of heat shock gene *hsp*70 from Drosophila, was introduced into tobacco, induction of NPTII expression due to heat shock was comparable to that of endogenous plant heat shock genes (inducible in all organs except pollen). Similarly, in maize a homologue of above gene, *hsp*70 with 1.1 kilobases of upstream sequence, when introduced in transgenic petunia, exhibited heat inducibility.

Transgenes that are inducible (by a stimulus) and/or exhibit tissue-specific expression in transgenic plants.

Gene product	Donor plant	Transgenic host	Inducer stimulus	Specific host tissue
1. rbc S	Pea	Tobacco, petunia	Light	Green tissues
rbc S	*N. plumbaginifolia*	N. Plumbaginifolia, petunia, tobacco	Light	Green tissues
rbc S	Soybean	Petunia	Light	Green tissues
2. Cab	Pea, wheat	Tobacco	Light	Green tissues
Cab	*N. plumbaginifolia*	Tobacco	Light	Green tissues
Cab	Wheat	Tobacco	Circadian clock	Green tissues
3. ST-LSI	Potato	Tobacco	Light	Green tissues
4. Chalcone synthase	*Antirrhinum majus*	Tobacco	Light	Flower
5. Hsp	Maize	Petunia	Heat	All tissues
	Soybean	Tobacco	Heat	All tissues
6. Protease inhibitor	Potato	Tobacco	Wounding	All tissues
7. Patatin	Potato	Potato	Tuber growth	Tuber
8. Leghaemoglobin	Soybean	Lotus	Rhizobium	Nodule
9. Nodulin	Soybean	Lotus	Rhizobium	Nodule

Contd...

Gene product	*Donor plant*	*Transgenic host*	*Inducer stimulus*	*Specific host tissue*
10. β-phaseolin	French bean	Tobacco	Rhizobium	Seed
11. β-conglycin	Soybean	Petunia, tobacco	Seed development	Seed
12. Lectin	Bean	Tobacco	Seed development	Seed
	Soybean	Tobacco	Seed development	Seed + root
13. Glutenin	Wheat	Tobacco	Seed development	Seed
14. Hordein	Barley	Tobacco	Seed development	Seed
15. Legumin	Pea	*N. plumbaginifolia*	Seed development	Seed
16. EPSP synthase	Petunia	Tobacco	Herbicide	Flower pollen
17. Ribonuclease	1. *Aspergillus oryzae*	Tobacco	Anther development	Anthers
	2. *Bacillus amyloliquefaciens*	Tobacco	Anther development	Anthers
18. Ribonuclease inhibitor	*Brassica amyloliquefaciens*	Tobacco	Anther development	Phloem cells

Alcohol Dehydrogenase Gene (ADH) : Alcohol dehydrogenase gene expresses under anaerobic conditions involving oxygen deprivation. Using transgenic plants, it could be shown that regulatory sequences upstream to ADH gene from maize lead to suppression (downregulation) of gene activity under aerobic conditions, this activity being normal and uninducible under anaerobic conditions. The inability of the ADH upstream regulatory sequences to induce gene expression became evident when these sequences failed to induce chloramphenicol acetyltransferase (CAT) expression under aerobic as well as anaerobic conditions in transgenic tobacco plants. Other enhancer sequences (from octopine synthase or CaMV35S gene), when transferred at a position upstream of ADH gene, led to induction under anaerobic conditions. Therefore, under normal conditions, ADH activity is inhibited under aerobic conditions due to negative control and expression is promoted under anaerobic conditions simply due to the absence of a negative control rather than the presence of positive induction.

Genes for Seed Storage Proteins in Legumes and Cereals: The genes for seed storage proteins provide an excellent example of a cell and tissue-specific expression. Using transgenic plants, sequences could be identified which were responsible for tissue or cell-specific expression of these genes. For instance, when wheat glutenin gene or barley hordein gene (with its regulatory sequence) was fused to chloramphenicol acetyltransferase (CAT) gene and introduced into tobacco, CAT activity was found only in the endosperm. Another reporter gene utilized for the same purpose is p-glucuronidase (GUS) gene, which also showed endosperm- specific expression when fused with a wheat glutenin gene (with its regulatory sequence). The regulatory sequence thus identified was called endosperm box of the storage protein genes.

EPSP Synthase Gene : 5-enopyruvyl shikimate-3-phosphate synthase (EPSPS) is an enzyme involved in aromatic amino acid biosynthetic pathway and is the target of lyphosate-based herbicides. In petunia, enzyme is synthesized in flowers at a high level and is hardly found in leaves, although in tomato it is synthesized both in flowers and leaves. Transgenic tobacco plants carrying CAT gene or GUS gene fused with deletion mutant forms of EPSPS gene were analysed and it was concluded that sequences between -1800 to -800 of petunia EPSPS gene were responsible for tissue-specific expression of this gene.

Transforming Process of DNA Sequences/transcripts and Histone Proteins : Regulation of gene expression can also be achieved through methylation and demethylation of coding DNA sequences or their transcripts. Histone proteins involved in the organization of nucleosomes can also be modified (acetylated) to allow access of transcription machinery to the DNA sequences, thus regulating gene expression.

DNA/RNA Methylation Sequences : During 1980's, a definite role of methylation and demethylation in gene control was also established. It was shown in many genes that methylation of specific sites in the vicinity of a gene may lead to suppression of gene activity and demethylation may lead to expression of gene. With the help of restriction enzymes like Hpall and Mspl (they are called isoschizomers i.e. they recognize same restriction sequence, but one of them does not cleave it when methylated), degree of methylation at CpG (cytosine - guanine) rich sites could be studied to show that the sites are under methylated in tissues, where the gene is active. Evidence of suppression of gene activity due to methylation is available from several experiments, including the following: (i) Herpes thymidine kinase (*tk*) gene was first methylated in vitro and then used for transformation of mouse tk^- cells. It was found that efficiency of transformation was low and the few transformed tk^+ cells available had partially demethylated DNA in comparison to the methylated tk gene used for transformation. Similar results for another gene apart^{+} in mouse were also available, (ii) In mouse retroviruses, using transfection experiments, it was shown that the inactive provirus sequences were heavily methylated, while active provirus sequences were undermethylated. It was also shown that non-methylated proviral sequences were infectious, and methylated sequences were not. (iii) In cultured mouse embryo cells, differentiation could be artificially induced by inhibiting methylation using a base analogue 5 azacytidine (5 azaC), which is specific in inhibiting DNA methylation in vivo. It was also shown in these differentiated embryo cells that the DNA actually became demethylated. Effect of 5azaC was also shown in avian retrovirus, where it induced gene expression associated with demethylation at an endogenous level. In human inactivated X-chromosome in mouse-human cell hybrids also, 5azaC enhanced reactivation, presumably due to demethylation.

The above evidences suggest and support the model that absence of methylation is necessary for gene expression in many cases. It is believed that methylation or demethylation takes place at specific critical sites, which represent a small fraction of sites available in DNA. This results in alteration of a site in such a manner that a particular regulating protein may interact with the site in a very specific manner, thus regulating gene activity.

Acetylation of Nucleosome Histones and Regulation of Gene Expression : It has been shown that acetylation of histones can destabilize nucleosomes and allow access of transcription factors to recognition elements, relieving the gene sequences of repression due to histones. Conversely, deacetylation of histones stabilizes the repressed state. Transcriptional coactivators are now known, which have histone acetyl transferase activity. CBP/p300 is one such coactivator having acetyl transferase activity, integrating transcriptional-activation signals from diverse regulatory proteins including steroid and nuclear-hormone receptors. In addition of a ligand (like oestrogen and thyroid hormone), the steroid and nuclear hormone receptors stimulate CBP/p300 mediated histone acetylation, thus rendering the chromatin much more accessible to transcriptional machinery.

Alternative Splicing of Transcripts : Another regulatory mechanism in eukaryotes involves alternative splicing described in before hand. The expression of a large number of genes has been shown to be regulated in time and space by this mechanism. In this mechanism, the splice sites selected by splicing machinery depend on the conditions demanding differential expression of genes.

Translation-level Regulation

Information about translational control in eukaryotes is coming from experiments designed to answer following questions regarding the mechanism of operation of translational machinery. (i) When there are more than one AUG codons in a mRNA molecule, how does the ribosome know, which one to use for initiation of translation ? (ii) What determines the frequency with which a particular mRNA message is read ? (iii) After finishing translation of an open reading frame (ORF), can a ribosome start again, if an AUG codon follows in the same mRNA within a reasonable length ? (iv) Can a mRNA remain untranslated in a cell and translated only when necessary?

It has been possible to determine the nucleotide sequence of natural mRNA, to alter these mRNAs as desired and to make synthetic mRNA using recombinant DNA methods. These techniques helped in answering some of the above questions and also yielded information regarding regulation of gene expression at the level of translation. While describing Kozak's scanning hypothesis for reading of mRNA by ribosomes in eukaryotes.

Activation and Repression of Translation Emploting mRNA Binding Proteins : We described mechanisms of repression as well as activation of translation in prokaryotes. Several proteins involved in repression and activation of translation are thus known in prokaryotes. However, in eukaryotes, no activator protein is known, and there is only one documented example of a translational repressor protein (iron-responsive element-binding protein = IRE-BP). On the other hand, regulation of translation in eukaryotes is also brought about by modification of general components (e.g. eIF2) of the translational machinery, a phenomenon unknown in prokaryotes. Regulation is also exercised during elongation of polypeptide chain, although its mechanism is not well understood.

Activator Proteins : In eukaryotes, till recently, no unambiguous examples were available, where a m-RNA binding protein directly activates translation. However, one possible mechanism that may be involved in the action of an activation protein is shown in Figure. In some cases, the non-coding leader sequence at the 5' end forms a secondary structure, so that the 5' end is not available for binding of 40S subunit of ribosome.

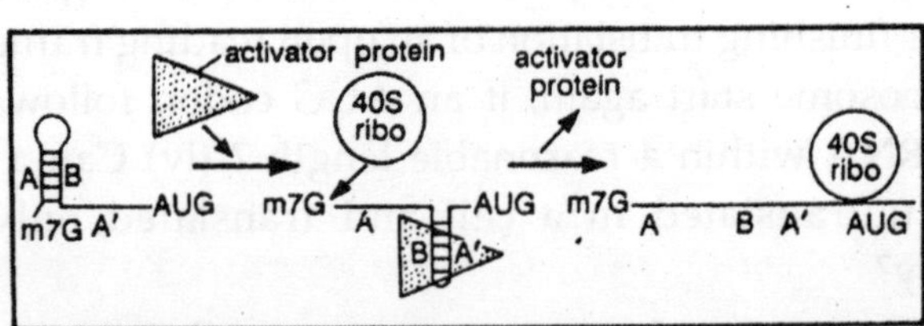

Hypothetical mechanism for activation of translation in mammalian systems, (a) In the absence of activator, 5' end acquires a secondary structure, which does not allow binding of 40S; (b) binding of activator (triangle) opens the 5' end, by making an alternative stem-loop structure; (c) the activator is displaced by processivity of scanning 40S subunit of ribosome.

The activator protein binds to mRNA and leads to the formation of a hairpin structure, which is accompanied with the exposure of 5' end of mRNA for ribosome binding. The activator protein is later displaced by the scanning 40S subunit of ribosome, till it comes across the initiating codon AUG.

For transcription, many gene-specific initiation factors (transcription factors) are known. For translation also, general initiation factors are known both in prokaryotes and eukaryotes, but no gene-specific or m-RNA specific translation initiation factors have yet been identified in eukaryotes (although for translation of bacterial mRNA and mt mRNA, such factors are known).

Repressor Protein, IRE-BP

(a) *Down-regulation of Ferritin Synthesis.* The synthesis of ferritin (an iron-sequestering protein) in mammalian cells depends on iron availability. Its regulation is achieved through inhibition of the translation of ferritin mRNA, due to binding of a repressor protein at the 5' end of this mRNA. This 5' end contains the iron-response element (IRE), which makes a stem loop structure. The repressor protein is similarly called iron-response element binding protein (IRE-BP). IRE-BP resembles mitochondrial aconitase and has aconitase activity, enabling it to sense iron levels. When iron is scarce, the reduced form (3Fe-4S) of the protein binds the IRE and when iron is abundant, the oxidized form (4Fe-4S) dissociates from IRE. The stem-loop structure of IRE, located +10 to +30 nucleotides from 5' end of mRNA, allows binding of IRE-BP, inhibiting the initial binding of 40S subunit of ribosome. This reduces or stops the synthesis of ferritin from its mRNA (Fig.).

(b) *Up-regulation of Transferrin Receptor (TfR) Synthesis :* The IRE-BP, while represses the synthesis of ferritin, it upregulates the synthesis of transferrin receptor *(TfR)*. The 3' end of TfR mRNA is prone to cleavage and IRE-BP binds there giving stability to TfR mRNA, so that it can be efficiently translated into TfR (Fig.).

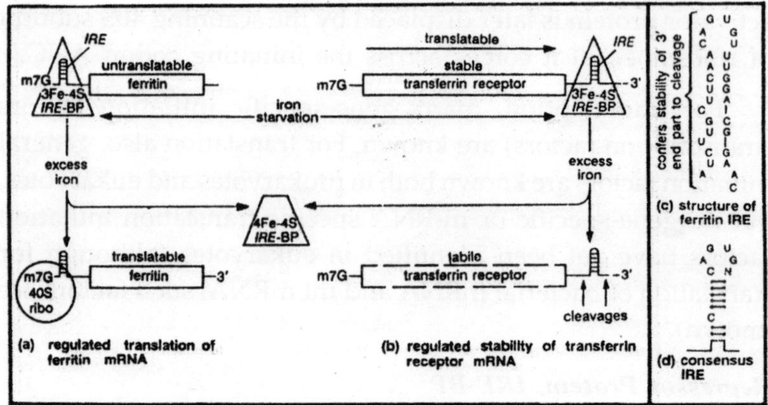

Post-transcriptional regulation of two genes involved in the uptake and detoxification of iron in mammalian cells, (a) translation of mRNA for ferritin, a protein that sequesters excess iron, is inhibited by the binding of IRE-BP to mRNA at its 5' end; (b) translation of transferrin receptor (TfR) involved in the binding of IRE-BP to TfR mRNA at its 3' end; (c) actual sequence of one IRE (TfR mRNA has five such IRE motifs at 3' end; only one is shown here); (d) consensus IRE structure (see text for details).

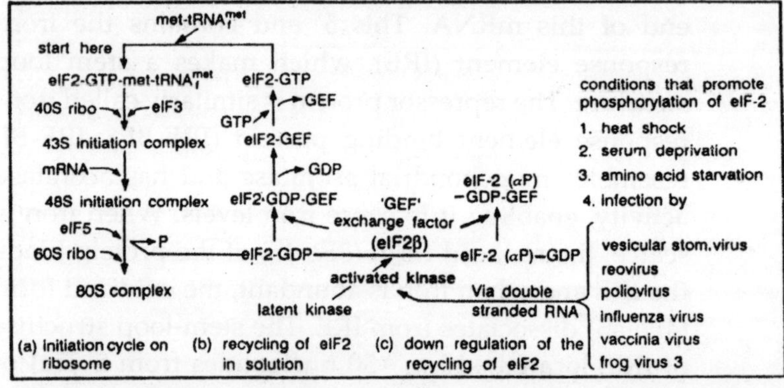

The eIF-2 cycle in mammalian cells. The eIF-GDP complex released from the 40S ribosomal subunit in step (a is stable. The exchange of GDP for GTP in step (b) is mediated by a protein, designated GEF which is usually limiting in cells.. Because GEF has a much higher affinity for eIF-2 (aP) than for unmodified elf-2, phosphorylation of eIF-2 sequesters GEF in a stable complex (step c), thus causing the global translational capacity to decline. The modification of eIF-2 and consequent inhibition of translation are reversed by phosphatases, which are not depicted. The enzymes that catalyze phosphorylation of eIF-2 include p68-kinase (DAI) and the heme-regulated kinase HRI. The increase in phosphorylation depicted in step (c) is itself subject to down-regulation, for example, by VA-RNA of adenovirus. The other viruses marked with an asterisk also appear to encode or induce components that circumvent the virus-induced phosphorylation of eIF-2 (see text).

IRE-BP also regulates translation of the first enzyme of heme biosynthetic pathway meant for iron utilisation. Thus a single repressor protein (IRE-BP) regulates both iron homeostasis and a metabolic pathway for iron utilization.

***Translational Control by Phosphorylation of the Elements of Translation Machinery* :** Several eukaryotic initiation factors (elFs) involved in the translation of mRNA, undergo reversible phosphorylation and are thus involved in the regulation of gene expression. The most important of these initiation factors are eIF2 and eIF-4E and their role in regulation of gene expression by reversible phosphorylation has been elucidated.

***Reversible Phosphorylation of eIF-2*α :** In mammalian systems, reversible phosphorylation of eIF-2a has been known for years, although its role in regulation of translation has been proved only recently. Such translational control by eIF2oc phosphorylation is not known in plants. eIF-2 forms a complex with GTP, which binds with met-tRNAmet and carries it onto the 40S ribosome subunit, forming a 43S complex. The 43S complex engages at the 5' end of mRNA and scans down to the AUG codon, where eIF-5 joins and causes hydrolysis of GTP. The resulting eIF-2-GDP has lower affinity for 40S and is therefore released, leaving 40S $tRNA_i^{met}$ behind. The 60S then joins giving 80S ribosome, ready to form the first peptide bond.

Affinity of eIF-2 for GDP is 400-fold higher than for GTP and therefore regeneration of eIF-2 GTP from eIF-2-GDP requires a helper factor, designated eIF-2β or GEF (guanine exchange factor). (In prokaryotes the function of GEF is performed by EF-Ts). However, the regeneration of eIF-2-GTP may be prevented, if α subunit of eIF-2 undergoes phosphorylation, which is promoted by a variety of conditions including heat shock, amino acid deprivation or virus infection. GEF has a 150-fold higher affinity for eIF-2(α P) than for unmodified eIF-2, so that formation of stable eIF-2 (α P) GDP-GEF does not allow the formation of eIF-2-GTP, causing protein synthesis (translation) to stop (Fig.). An example of eIF-2 mediated regulation of the translation of mRNA for the yeast gene GCN4 is discussed below.

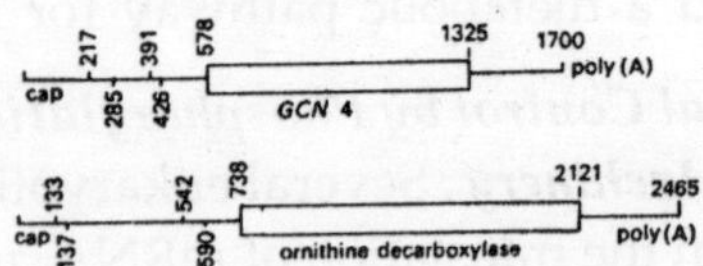

Structures of mRNAs corresponding to genes for GCN4 (general control of nitrogen metabolism) and ornithine decarboxylase in yeast, showing untranslated sequence at 5′ end having false AUG codons.

(a) *Translational Control of the Gene GCN4 in Yeast via eIF-2* : In yeast, expression of GCN4 gene (involved in general control of nitrogen metabolism) is regulated primarily at translation level, although under different conditions, the level of protein encoded by GCN4 changes more than 100 fold, the level of mRNA for this gene does not change. GCN4 gene protein (a transcription factor) regulates transcription of 30—40 gents involved in amino acid biosynthesis and its own synthesis is boosted 100 fold when yeast is starved of an essential amino acid. Its level of expression also depends on a repressor specified by GCD1 locus, which in its turn is regulated by GCN1, 2 and 3 loci.

The structure of mRNA of GCN4 gene explains the mechanism of its translational regulation on the basis of Kozak's scanning hypothesis. It has a 577 nucleotides long untranslated sequence at 5' end, which is much longer than normal and has four AUG codons upstream of the real AUG (Fig.). The presence of four AUG codons is utilized by the eukaryotic initiation factor, eIF2 in the regulation of translation of GCW4-mRNA. The initiation factor, eIF-2 forms a complex eIF-2-GTP-met-tRNA, whose concentration is regulated by amino acid concentration. Under amino acid starvation, uncharged $tRNA_i^{met}$ accumulates and activates a protein kinase, which phosphorylates and inactivates a portion of eIF-2 molecule, thus reducing the concentration of eIF-2 in the cell. Since eIF-2 escorts met-tRNA^{met} onto ribosomes, under reduced concentration of eIF-2, 40S subunit will require more time to acquire met-tRNA^{met} and become competent for reinitiation. Due to slower acquisition of competence, some 40S subunits bypass one or more AUG codons and gain access to the GCN4 start site. On the other hand, under condition of no starvation (amino acid), eIF-2 concentration is adequate so that the 40S subunits become competent quickly and cannot bypass the fourth AUG codon. Consequently, a small peptide is synthesized and released at the termination site of ORF4 (open reading frame-4). The termination at ORF4 is somehow incompatible with reinitiation at GCN4, so that under condition of no starvation, GCN4 is not expressed. Translation of GCN4-mRNA is also inhibited, if intercistronic distance between ORF1 and ORF4 is increased by inserting an unknown sequence. In this case, a longer distance allows all 40S subunits become competent before reaching ORF4, even under starvation condition. This will not leave any 40S subunit to bypass ORF4 and reinitiate translation at GCN4. (Fig.).

(b) *Translational Control of the Synthesis of Ornithine Decarboxylase:* An example analogous to GCN4 gene in yeast, is the mRNA corresponding to ornithine decarboxylase enzyme involved in polyamine biosynthesis. It contains a 737 nucleotides long 5' untranslated region containing four AUG codons (Fig.) followed by termination codons 16, 2, 4 and 10 nucleotides later respectively. The enzyme is extremely rapidly produced in cells stimulated by hormones and growth factors, but we do not know if translational efficiency changes with the physiological state of the cell. If so, the long leader sequence and the false AUG codons may play a part.

Reversible Phosphorylation of eIF-4E and rpS6 (Ribosomal Protein S6). Besides eIF-2, eIF-4E and rpS6 have also been observed to undergo striking changes in the level of phosphorylation under specific conditions like serum deprivation in culture medium (eIF-4E) and heat shock. Under heat shock, phosphorylation of eIF-2 increases and that of rpS6 decreases. Also, while increased phosphorylation of eIF-2 leads to reduced translation, increased phosphorylation of rpS6 apparently leads to enhanced translation.

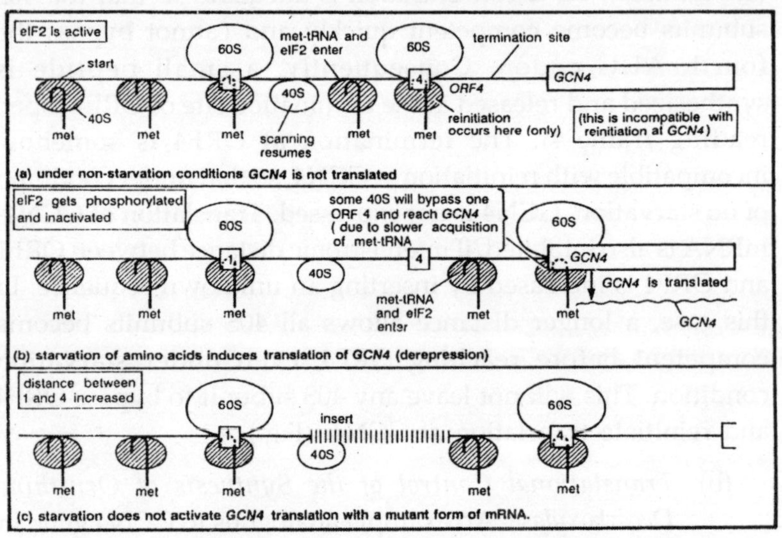

Translation control of GCN4 gene in yeast through reinitiation mechanism; only two of the four AUG codons (shown as 1 and 4) are depicted (see text for details).

In case of eIF-4E, the principal site of phosphorylation is serine-53′ and a mutation on this site renders the factor non-functional in translation assays. This suggests that eIF-4E may undergo reversible phosphorylation each time it is used. In activated B lymphocytes also, increased phosphorylation was associated with increased translation rate.

One of the cell division cycle (cdc) genes in yeast also encodes eIF-4E, which may regulate cell division by reversible phosphorylation. Phosphorylation of eIF-2, eIF-4E and rpS6 is also associated with phosphorylation of other initiation factors, elongation factors and aminoacyl tRNA synthetases. However, unequivocal evidence of the regulation of translation by reversible phosphorylation is available only in case of eIF-2.

Phosphorylation of Reversible Type and Targeting Specific mRNA : The regulation at the translation level requires that the phosphorylated or dephosphorylated forms of the components of translational machinery, should identify a specific mRNA from the bulk mRNA population. This may be achieved in one of the following ways : (i) a mechanism like that for GCN4 described abovc (several AUG codons upstream of real start codon); (ii) activation of kinases or phosphatases in a localized region of cytosol, where a specific mRNA is localized (compartmentalization of the cytosol); (iii) variation in secondary and tertiary structures among the non-coding leader sequences at the 5' ends of mRNA molecules, which modulate ribosome-mRNA interaction.

The best example of differential translation of mRNAs is the virus infected cells, where viral genes are preferentially expressed. This may be achieved by one of the following mechanisms : (i) Cellular mRNAs are degraded or host's transcription is shut off. (ii) Viral mRNAs are more abundant (e.g. reovirus). (iii) Viral mRNAs are better designed for initiating translation, due to unstructured 5' leader sequence, (e.g. adenovirus). (iv) Overall translational capacity of the host is reduced due to partial phosphorylation of eIF-2 α , induced by double-stranded RNA generated during the course of virus transcription and/or translation (e.g. adenovirus, poliovirus, reovirus, etc.). In many viruses, factors are also generated which inhibit completion of phosphorylation of eIF-2 α . An adenovirus encoded transcript, designated VA-RNA, actually binds to and blocks the activation of p68 kinase thus making unaltered eIF-2a available for efficient translation.

Masked mRNA in Eggs of Sea Urchin and Xenopus

There are also several examples of 'masked mRNA' in eggs and oocytes, which may indicate translational control. This 'masked mRNA' is found in abundant quantity and remains untranslated until after fertilization. There are at least two examples of this 'masked mRNA' :

(i) the mRNA coding for small subunit of ribonucleotide reductase in clam and sea urchin eggs and

(ii) the mRNA for fibronectin in Xenopus unfertilized eggs, remain untranslated until about 6-7 hours after fertilization (midblastula stage). The mechanism of this unmasking is not clearly understood.

Gene Rearrangement Regulation

Immunoglobulin Genes Expression : Antibody molecules, called immunoglobulins, are coded by genes, which undergo DNA rearrangements during cell differentiation, thus suggesting that although a genome usually remains constant in different cells, it does undergo a change with regard to immunoglobulin genes in mammals. Immunoglobulins are produced by cells originating in bone marrow (B-cells), and later differentiating into lymphocytes. These may differentiate into several clones, producing different antibodies, but each clone producing only one antibody. For this work on the rearrangement of DNA sequences that results in a large variety of sequences for production of antibodies, S. Tonegawa of MIT (USA) was awarded the 1987 Nobel Prize for Physiology/ Medicine. This work will be briefly described in this section.

Immunoglobulin Chains Structure : Antibody molecules (immunoglobulins, abbreviated as I_g) are tetrameric, each containing two identical 'heavy' (H) chains of about 500 amino acid residues, and two identical light (L) chains of about 250 amino acid residues. In man, mouse, rabbit and other mammals

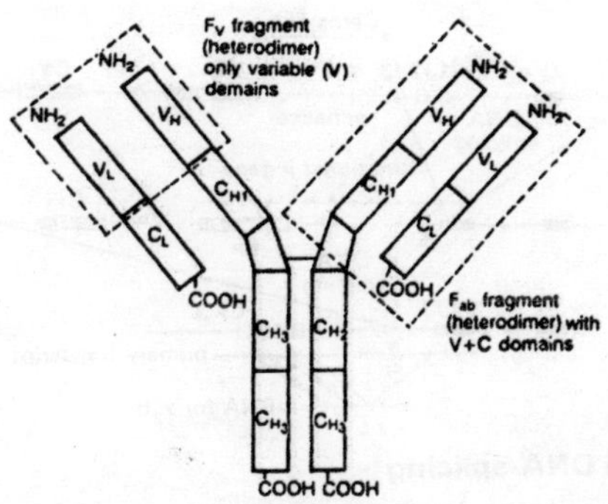

Heavy and light chains and their association to generate different V (variable) domains and C (constant) domains in immunoglobulin protein.

there are seven kinds of heavy chains (alpha = α ; gamma = γ_1, γ_{2a}, γ_{2b}, γ_3; delta = δ ; mu = μ and two kinds of light chains (kappa = K and, lambda = λ). I molecules are named according to the type of H chain, which they contain (e.g. I_gA has gamma chains, I_gM has mu chains and I_gA has alpha chains).

Each I_g chain consists of two distinct parts (i) C-terminal constant part or C-region and (ii) N-terminal extremely variable part or V-region. For each C-region (C_α, C_μ, C_λ, etc.), there can be several V sequences. Separate sets of V sequences are known for heavy chains (V_H) and light chains (V_κ - for kappa and V_λ for lambda). The V_H, V_κ and V_λ sequences have fair degree of homology and possess only some variable regions.

I_g gene in Germline and Differentiated Cells : By using cDNA as probe, it could be shown that C-and V-coding sequences were closer to each other in the antibody-producing cells, than in the germ line. This was predicted because, same C-region was found combined with different V-regions in the same individual. Later experiments showed that C—C splicing at the DNA level is responsible for generating the coding genes for both light and heavy chains. Although, the genes for immunoglobulin molecule may be found on more than one chromosome, but cutting and rejoining of DNA takes place

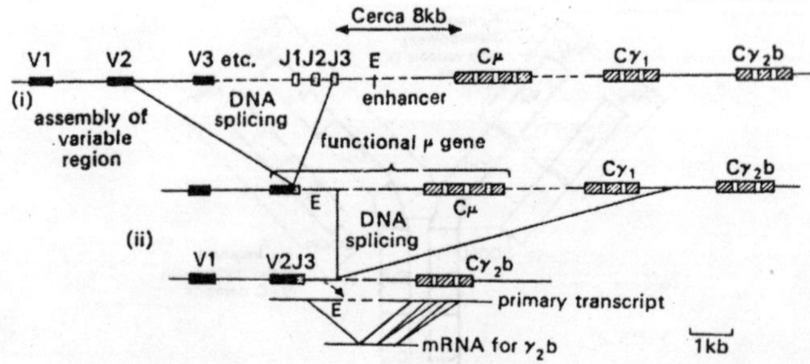

(A) steps in DNA-splicing

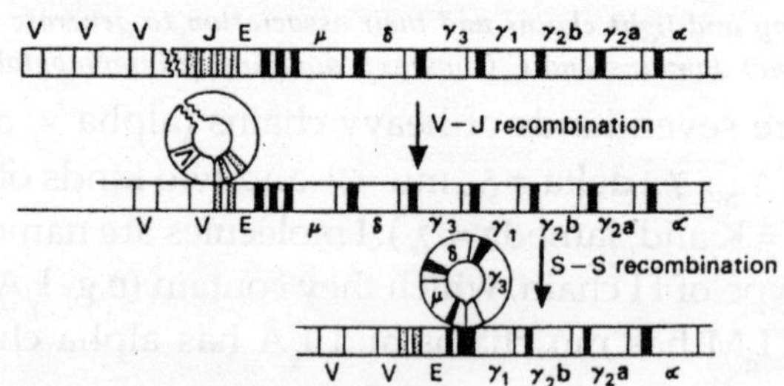

(B) model of heavy-chain switching

(A) structure of immunoglobulin gene and the steps involved in DNA rearrangements in different regions. (B) a model explaining DNA rearrangements in heavy chain switching.

only within a chromosome and not across chromosomes. In addition to C and V regions, both light and heavy chains have another region called J (junction of joining sequences). In heavy chain, in addition to C, J and V regions, another short sequence D (diversity) is also found.

We will discuss DNA splicing only for heavy chain as an example. There are following two rearrangements in this case; (i) the first generating the DNA sequence coding for V_H, and (ii) the second leading to association of specific V_H with a definite C_H sequence. Using probes in different kinds of cells, steps involved in rearrangements leading to association of different V regions could be worked out. In LM sequence, it is shown that in embryo or germ line cells, V_H and C_H regions have a small region of J sequence. Between V and J sequences, presence of D sequence was also shown.

V-regions Through DNA Splicing and Joining : Each heavy chain V_H region is encoded by a DNA sequence assembled from three gene segments a V segment, a J segment and D segment. The antibody diversity in the V*sub* region is based on (i) inherent diversity in the sequences representing V, J and D segments and on (ii) the combinatorial joining of these segments (called combinatorial diversity). Any of the 500 V segments in the mouse heavy-chain pool can be joined to any of the 4 J segments and to any of at least 12 D segments, to encode at least 24,000 (500 × 4 × 12) different, heavy chain V_H regions. Similarly, in the light chain at least 1000 V_L regions can be obtained, so that there can be as many as 24,000 × 1000 = 25×10^6 divergent antibodies. The number of possibilities are further increased by the joining, mechanism itself (there may be loss or gain of nucleotides during joining). This random loss or gain at joining sites is described as junctional diversity. Two genes (rag-1 and rag-2; rag = recombination activating genes) are involved in the U(D)J recombination system to produce heavy chain V_H regions.

V-C association Through DNA Splicing : Second phase of DNA splicing involves association of a V segment with C segment. It has been shown that the same V segment can attach with different C-regions, thus providing for H-chain class switching. This is achieved by successive deletions of DNA, leading to the production of I M in primary response and I G and I_gA as secondary responses. The V-D-J segment may first attach to C_μ (which is first exon in C-region) and give I_gM. Later the cell may switch and may produce another immunoglobulin. This class switch may require further deletions of C-region exons. The deletion is mediated by a repeated DNA segment located just before each C-exon and designated as S-region (switch). I M has all kinds of heavy chains; in IgG_3, segment C is missing and in $I_g G_1$, C_μ and $C_{4\gamma}^{3}$

are missing. Further deletions may occur in other cases as shown in Figure.

Enhancers in Regulation of Immunoglobulin Genes. Enhancers are certain DNA sequences, that increase the transcription of genes (see page 609), although this is not the only characteristic of DNA segment to be called enhancer. Other essential characteristics of an enhancer are : (i) it is essential for transcription; (ii) it can be moved around on either side of gene or may be inverted and (iii) it can work with genes from sources other than its own. These enhancers have been discovered in immunoglobulin genes and have been found to work only in cells that make immunoglobulins. In this manner, enhancer is a remarkable regulatory sequence that is cell specific. The enhancer sequence is found (i) between V-D-J and C in DNA segment for heavy chain and (ii) between V-J and C in DNA segment for light chain (Fig.). The role of enhancer sequence is demonstrated by the following evidences, (i) If intron region between V-D-J and C regions of heavy chain is removed from a cloned gene system, and then the latter is injected into mouse myeloma cells, the cloned genes lose their ability to make a heavy chain, (ii) When cloned intron region was cleaved into three fragments, and each is separately linked with genes for T-antigen of SV40, or with (3 globin gene of mammals, only one of these fragments, presumably carrying the enhancer, stimulated the transcription, (iii) When cloned intron region of heavy chain gene was attached to a bacterial gene, increased gene product was achieved.

To obtain correct initiation of transcription, the enhancer must interact with other control sequences which are located upstream (towards 5' end) from the beginning of V-region. Normally V-D-J (or V-J) joining the upstream sequence and enhancer are far apart, but rearrangements bring them close enough for transcription to occur. The enhancer functions even after every class switch and thus can continue to activate transcription.

Type of Yeast Mating Switching : Life cycle of budding yeast, S. cerevisiae (Fig.) shows that there are two mating types (MATa/MAT α), which fuse to give a diploid (MATa/MAT α).

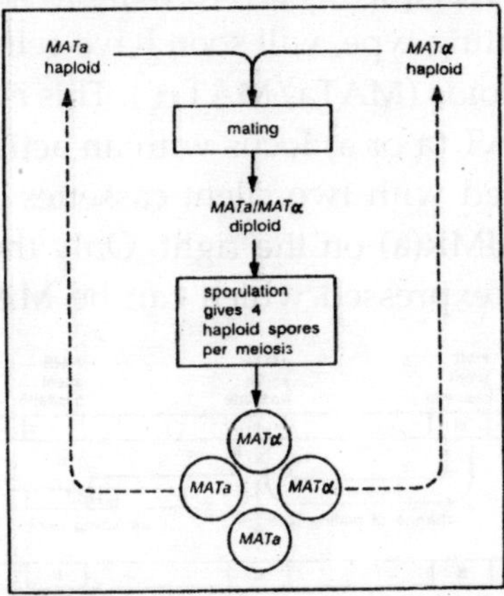

Life cycle of yeast involving mating of MATa and MAT α haploids.

The diploid cell sporulates and through meiosis forms four spores, two of each mating type. Each mating type is characterized by secretion of a pheromone (MAT secretes a factor and MAT α secretes α factor) and a surface receptor for the pheromone of the opposite type. The diploids (MATa/ MAT α) produce no pheromone or surface receptor. The pheromone-receptor interaction activates a cell membrane protein called G protein consisting of three subunits (α β γ) (see later in this Chapter for details on G proteins; Nobel Prize for discovery of G protein was awarded to M. Rodbell and A. G. Gilman in the year 1994). Activation needs a GTP molecule, which displaces GDP associated with α β γ complex and releases α from β γ dimer. This separation of α from β γ activates the next protein in signal transduction pathway. Therefore G protein is also called a transducin. The monomer α alone (and not β γ) is capable of bringing about the activation. The above information has come from a study of mutants (called STE = sterile), which are incapable of mating. At least ten such mutations (STE 1-7, 11, 12, 18) are known.

A remarkable feature in yeast is the ability of some strains (which carry dominant *HO* allele and not those which carry recessive ho allele) to switch their mating type from a to α and vice versa. Consequently an *HO* haploid strain, irrespective of its original mating type, will soon have cells of both mating types giving diploids (MATa/MATα). This is achieved by the presence of a MAT (a or a) locus with an active cassette (type a or α) associated with two silent cassettes at loci HML(α) on the left and HMR(a) on the right. Only the active cassette at MAT locus is expressed which can be MATa or MATα.

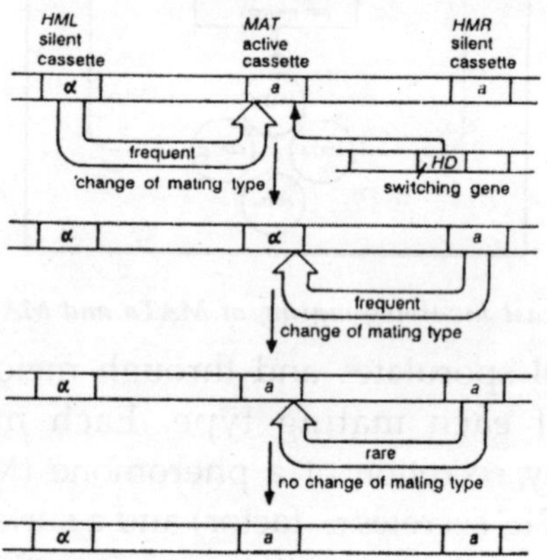

Change of mating types in yeast due to replacement I by active cassette by silent cassette of opposite type : when replacement involves cassettes of same type, the mating type remains unaltered.

This can be replaced by HMLa or HMRα respectively leading to change in mating type (α→α or α→α). MATα is almost always replaced by, HML and MATa by HMR, so that if HML(α) is mutated to HMLα or HMR is mutated to HMRα, replacement by them does not change mating type (Fig.). The *HML* and *HMR* are constitutively repressed (rendered silent) by some repressors binding to silencer sequences, HML-E or EL and HMR-E or ER (also called E and *I* silencers) located ~ 1 kb upstream of these loci (as discussed earlier). The switching is achieved by the activity of an endonuclease coded by the dominant gene HO. The gene HO is also repressed by several control systems, which need to be lifted before HO plays a role in switching.

Trypanosome surface antigen (VSG) switching. Sleeping sickness in humans (and a related disease in cow) is caused by a unicellular parasite called trypanosome, which alternates between two hosts, tsetse fly and a mammal (Fig.). Trypanosome undergoes important biochemical changes involving diversity in the variable surface glycoprotein (VSG), a major component of surface coat, which provides antigenic reaction. Although, at any one f time a trypanosome expresses only one VSG, its ability to change into any of the -100 possible different VSGs is the secret of parasite's' survival during *fly*-mammal cycle.

A tsetse fly, during a bite, gains the parasite, which loses its VSG but reacquires a new VSG after three weeks when, it differentiates in 'metacydic form'. This form enters mammalian bloodstream during a bite, where VSG keeps on changing every 1-2 weeks, so that the immune response always lags behind the change in VSG and the parasite evades the immune surveillance, perpetuating indefinitely, till it enters the central nervous system find often causes death.

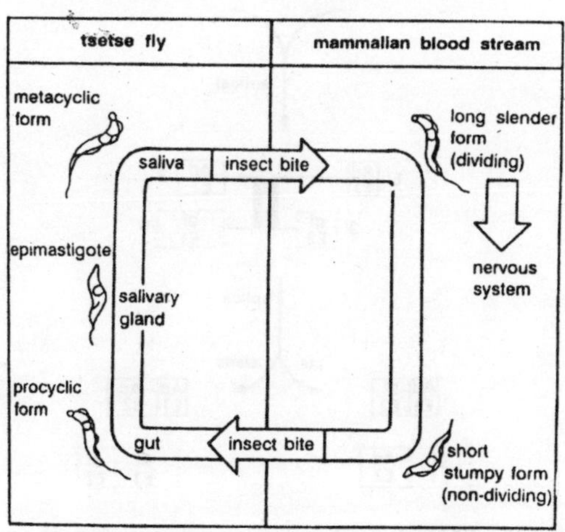

Life cycle of a trypanosome alternating between two hosts (tsetse fly and a mammal) causing sleeping sickness.

Genes (about 1000) for~100 different VSGs are present scattered on different chromosomes in the genome of trypanosome. Diversity, therefore, depends on changing expression from one pre-existing gene to another. Each VSG is coded by a single basic copy gene, which may be telomeric or internal in location, and there may be several isogenes for same VSG or similar VSGs. The copy of the gene, which is active, is called expression linked copy (ELC) and is located on an expression site. (Creation of an ELC may involve transfer of a basic copy gene to the expression site or vice versa. Almost all switches in VSG type involve replacement of the ELC by a pre-existing silent copy.

Synthesis of mRNA in Pieces in VSG Genes in Trypanosome: It is well established that in eukaryotes, a precursor mRNA is first synthesized, and is processed to give rise to mRNA. In split genes also, a pre-mRNA is first synthesized and intron regions are then excised to give rise to mRNA. However, in trypanosome (Trypanosoma brucei), it was shown that the

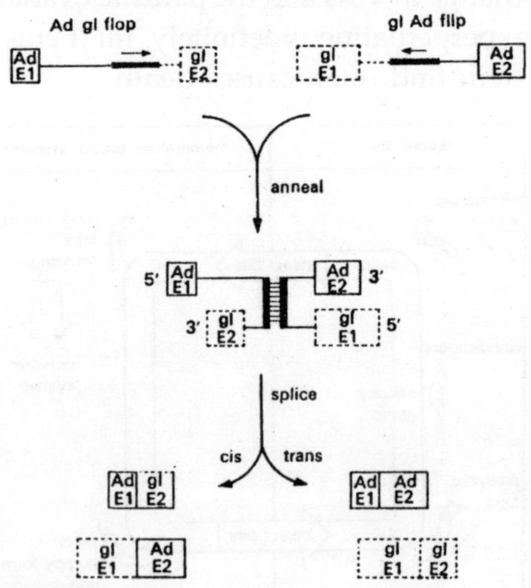

Production of mRNA through splicing and rejoining of segments from different precursor mRNA molecules. (For details, see text).

mRNA for some glycoproteins synthesized by VSG genes, has a small segment at its 5' end which is synthesized as a discrete transcript separate from the remainder of mRNA. It was also shown that DNA encoding this 35-nucleotides leader RNA sequence (called 'mini exon') of all VSG mRNAs is separated by 30 kilobases from the DNA segment coding for the remaining mRNA. This 35-nucleotides sequence is common to all trypanosome mRNAs. No precursor giant mRNA, which could give rise to this mRNA by post-transcriptional splicing could be observed. It was therefore, concluded that 35-nucleotides 'mini exon' RNA was synthesized as a part of a separate transcript and then either (i) joined to the remainder of the m-RNA through trans-splicing or (ii) acted as a primer for the transcription of the longer segment of mRNA.

It has also been shown that mRNA can be produced from two precursor molecules through trans-splicing followed by joining. This became possible through the availability of a technique for in vitro splicing systems. In a study from Massachusetts Institute of Technology, USA, two artificial mRNA precursors could be synthesized, one having LI exon and the other having L2 exon. These two exons could be spliced together under optimum in vitro conditions to enable them to join together. In another study from Yale University, USA, chimeric precursors had either (i) the exon adE_1 and glE_2 or (ii) the exons adE_2 and g1Ep. In each of these chimeric precursors, the two exons were connected by an intron and it was shown that splicing may be (i) cis leading to production of mRNA having exons from the same precursor or may be (ii) trans leading to the production of mRNA having exons from two precursors. The former is an intramolecular reaction and the latter is an intermolecular reaction and both were shown to proceed at the same rate.

In the above experiments, it was shown that the introns of two precursors have complementary sequences and, therefore, can be brought together either by base pairing or by mutual binding of a polypeptide. A scheme of cis and trans splicing of two chimeric precursor mRNAs to give rise to one mRNA is shown in Figure.

Gene Expression's Regulation by Reversible Protein Phosphorylation : The 1992 Nobel Prize for physiology or medicine was awarded to Prof. Edwin G. Krebs and Prof. Edmond H. Fisher, both of University of Washington, Seattle, U.S.A. for their pioneering work on 'reversible protein phosphorylation as a biological regulatory mechanism'. Phosphorylation of proteins has been shown to affect transcription, translation, cell division, cell growth, cell differentiation, cell transformation and many other cellular processes. Regulation of enzyme activity by reversible phosphorylation has several advantages over de novo synthesis of enzyme. The response is fast and the signal gets amplified. One molecule of kinase can phosphorylate several molecules of an enzyme. Activity of several non-enzymatic proteins like transcription factors is also regulated by reversible phosphorylation. It is estimated that a eukaryotic cell has about 1000 genes for different protein kinases (for phosphorylation) and perhaps same number of genes for protein phosphatases (for dephosphorylation).

The first protein (enzyme in this case), whose activity was shown to be regulated by phosphorylation was glycogen phosphorylase (involved in breakdown of glycogen in muscle and liver). This enzyme exists in two forms, phosphorylase a

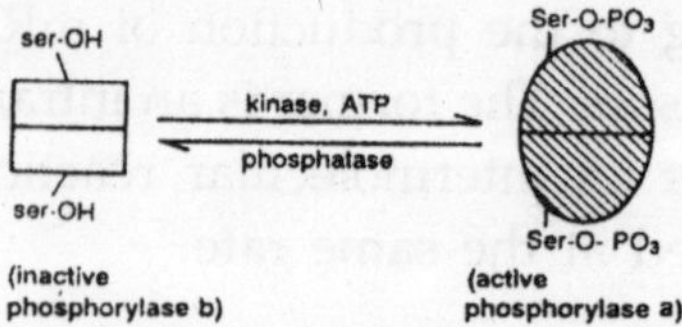

Reversible phosphorylation of phosphorylase enzyme at serine residue causing conformational changes in the enzyme.

(not dependent on AMP for its activity) and phosphorylase b (dependent on AMP for its activity). Reversible phosphorylation leading to interconversion of these two forms has been shown. Almost all of the enzyme is in an inactive 'b' form in resting muscle, which gets converted to the active 'a' form, when the level of hormone, epinephrine, increases in the blood (Fig.). Phosphorylation of protein occurs on many residues, but mainly serine (Ser; 90-95%), threonine (Thr; 5–10%) and tyrosine (Tyr; <10%) are phosphorylated. Phosphorylation does not always lead to increase in enzyme activity, and in some cases may actually lead to decrease in activity (e.g. glycogen synthase). Sometimes positive and negative regulatory phosphorylation sites are present in the same enzyme, so that same enzyme may need two kinases and two phosphatases.

Proteins of WD-repeat and their Regulatory Part : WD-repeat proteins are made up of conserved repeating units, each unit ending with Trp-Asp (WD). They are found in all eukaryotes, but not in prokaryotes. During early 1990's, this conserved repeat unit was found for the first time in beta (β) subunit of G-proteins, which transduce signals across the plasma membrane. Later, these WD repeats were found in more than two dozen proteins. Each repeat has a region of variable length, followed by a core of more or less constant length having GH (Gly-His) on one end, and WD (Trp-Asp) on the other end, so that the repeats can be represented as- $\{X_{6\text{-}94}\text{-}(GH\text{-}X_{23\text{-}41}\text{-}WD)\}_n$.

All WD-repeat proteins are regulatory in nature and none is an enzyme. They regulate cell functions, such as signal transduction (e.g. GB, I PR55, RACK1), RNA processing (e.g. CstF, Mak11, PRP4), gene regulation (e.g. COP1, $TAF_{II}80$, TUP1), vesicular traffic (e.g. β' COP, SEC 13) and cytoskeletal assembly/ cell cycle (e.g. CDC4, CDC20, β Trcp). Several of these WD proteins form multiprotein complexes, sometimes interacting with other proteins through the WD-repeat region.

Receptors of Cell, Cell Signaling and Regulation of Gene Expression : Since the regulation of gene expression should meet the requirements of a cell during different developmental stages and during adaptation to different environmental conditions, there should be mechanisms at the cellular level, which will perceive the extracellular and intracellular signals. Mechanisms should also be available to transmit the signal to the nucleus, for activating transcription of specific genes. These signals, when intercellular are mediated either through the secretion of signaling molecules or through membrane, bound cell surface molecules. The signaling molecules may include a variety of molecules including proteins, small peptides, amino acids, nucleotides, steroids, retinoids, fatty acid derivatives and even dissolved gases like nitric oxide and carbon monoxide. In each case, the signaling involves a ligand (the signaling molecule) and a receptor. The receptor can be an intracellular receptor or a cell surface receptor (Fig.).

The signaling molecules that are secreted by cells can mediate signals in one of the following four ways; (i) paracrine signaling involves signaling molecules that are secreted to

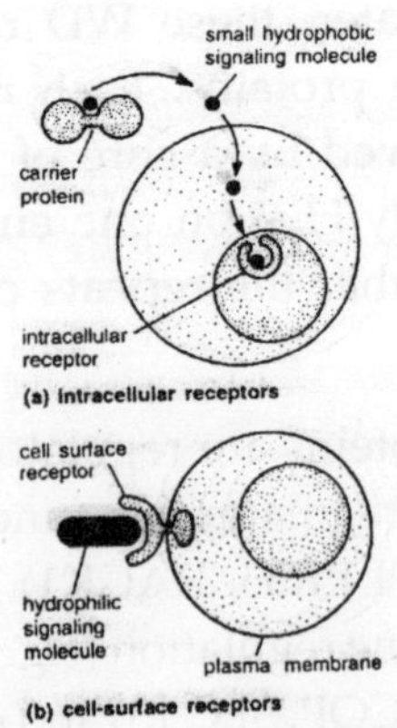

Intracellular receptors and cell surface receptors, each providing site for the binding of extracellular signaling molecule (ligand), which may be hydrophobic (can cross cell membrane) for intracellular receptors and hydrophilic (unable to cross cell membrane) for cell surface receptors (redrawn from Alberts et al., 'Molecular Biology of the Cell', 1994).

influence only cells in the vicinity; (ii) synaptic signaling involves the use of neurotransmitters, which transmit signals through specialized junctions called synapses found between nerve cells (consult a Cell Biology book), (iii) endocrine signaling involves the use of signaling molecules called hormones; (iv) autocrine signaling involves the secretion of molecules that bind receptors on the same cells or cells of the same type (as in contraction of smooth muscles or aggregation of platelets).

Intracellular Receptors : Steroid hormones, thyroid hormones, retinoids and vitamin D, when act as signaling molecules (ligands), diffuse across the plasma membrane of target cells and bind receptors available within these target cells. These ligands activate the intracellular receptors, which then regulate transcription of specific genes. These receptors are collectively called intracellular receptor superfamily (or steroid-hormone receptor surperfamily, Fig.).

Receptors of Cell Surface : The cell surface receptor proteins act as signal transducers. The ligand binds to a cell surface receptor with high affinity and activates the receptor, which converts the extracellular signal into one or more intracellular signals that alter the behaviour of the target cell. Cell surface receptors can be classified into following three groups (Fig.); (i) Ion channel-linked receptors (also called transmitter-gated ion channels). These are involved in rapid signaling at synapses (consult a book on 'Cell Biology'), (ii) G-protein linked receptors. These are involved in activating another membrane-bound target protein through a third protein called G protein (a trimeric GTP-binding regulatory protein). This target protein activated through G-protein, either alters the concentration of one or more intracellular mediators or modifies the ion permeability of the cell, leading to an alteration in the behaviour of the cell. All G-protein-linked receptors belong to a large surperfamily of seven-pass transmembrane proteins, (iii) Enzyme-linked receptors. These receptors, when activated, function directly as enzymes or are associated with enzymes (mostly kinases), thus bringing about phosphorylation of specific sets of proteins in the target cell.

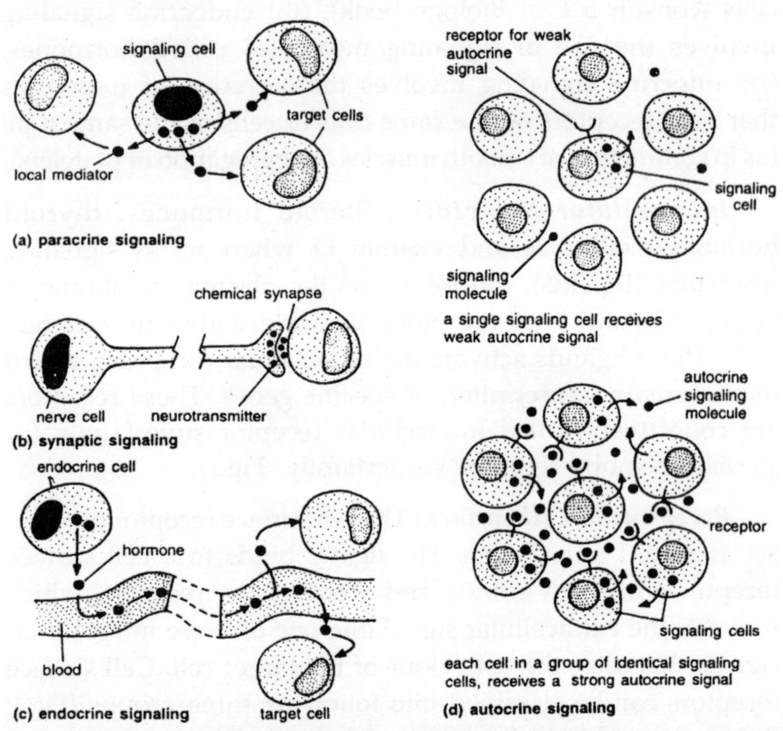

Four types of signaling mediated by secreted molecules, and differing in speed and selectivity with which the signals are delivered to their targets (redrawn from Alberts el ill., 'Molecular Biology of the Cell', 1994).

The ion-channel-linked receptors mentioned above are dealt in some detail in author's another book *"Cell and Molecular Biology"*. The other two types of receptors i.e. G-protein and enzyme linked receptors initiate a cascade of events. Each such event involves phosphorylation of an inactive protein, thus altering it into an active protein, which eventually leads to the expression of one or more specific genes. The phosphorylation cascades are facilitated by two main types of kinases. (i) serinelthreonine kinases , which phosphorylate proteins on serines and less often on threonines and (ii) tyrosine kinases, which phosphorylate proteins at tyrosines. Some enzymes

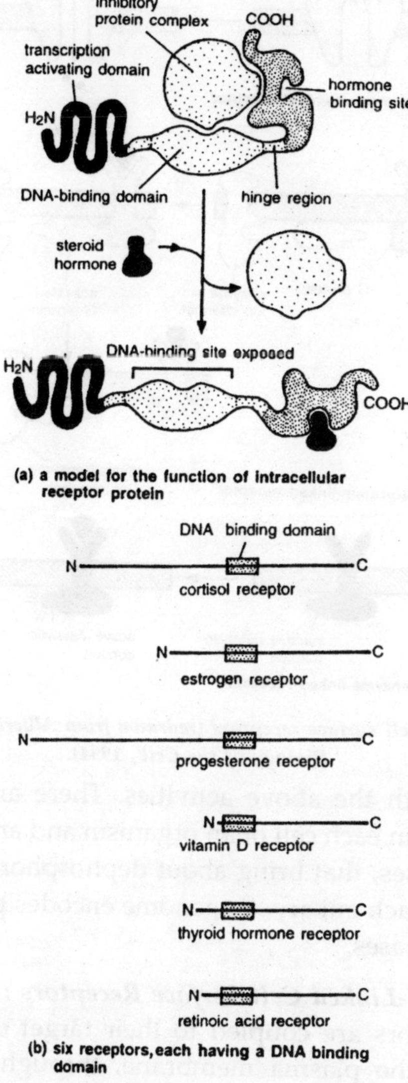

The intracellular receptor superfamily; (a) a model for the mechanism of action of an intracellular protein; (b) a variety of structures known in six different intracellular receptors (DNA- binding domain is shown by a box in each case) (redrawn from Alberts et al., 'Molecular Biology of the Cell' 1994).

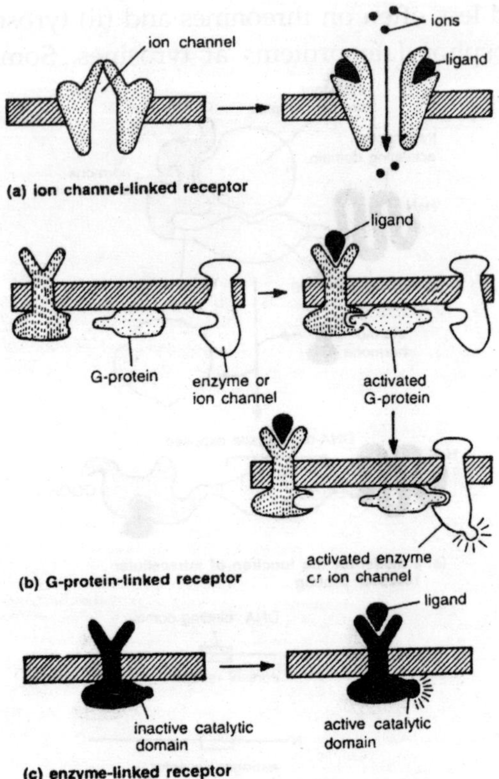

Three classes of cell surface receptors (redrawn from Alberts et at, 'Molecular Biology of the Cell', 1994).

may have both the above activities. There are hundreds of these kinases in each cell of an organism and an equal number of phosphatases, that bring about dephosphorylation, so that about 1% of each eukaryotic genome encodes protein kinases, and phosphatases.

G-protein-Linked Cell Surface Receptors : The G-protein-linked receptors are coupled to their target enzymes or ion channels in the plasma membrane, through trimeric GTP-binding proteins, called G-proteins. These trimeric G-proteins differ from monomeric GTP-binding proteins (often called GTPases), that relay intracellular signals and regulate vesicular traffic and many other processes. Actually both of these GTP-binding proteins are GTPases and are active, when bound to

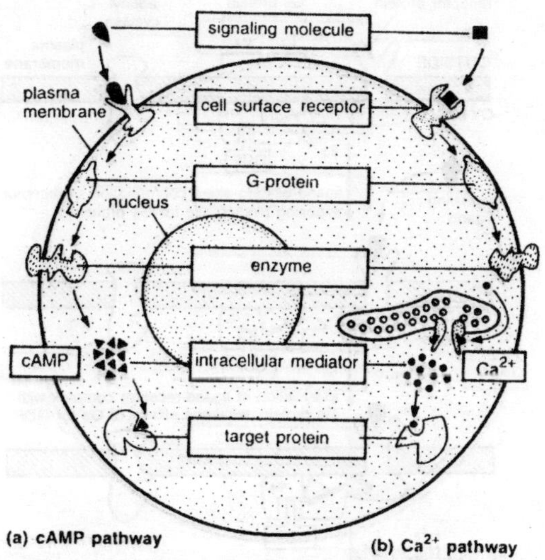

Two major pathways by which G-protein linked cell surface receptors initiate cell signaling pathways (redrawn from Alberts et al., 'Molecular Biology of the Cell', 1994).

GTP and inactive when bound to GDP. When a ligand binds to a G-protein linked receptor, the latter undergoes a conformational change, enabling the trimeric ($\alpha\beta\gamma$) G-protein (with associated GDP) to bind to this receptor. G-protein, in its turn, when associated with receptor, ejects its GDP and replaces it with GTP. This leads to dissociation of G-protein into two units $\beta\gamma$ and α (with GTP); the latter diffuses away from the receptor and delivers the message to the target enzyme or ion channel. The GTP is now hydrolysed and G-protein returns to its original trimeric form with associated GDP (Figs.). For the discovery of G-proteins, the Nobel Prize in Medicine for the year 1994 was awarded to Martin Rodbell and Arthur Gilman.

One of the most common enzymes activated by a class of G-proteins (stimulatory G protein or Gs) and inhibited by another class of G proteins (inhibitory G proteins or Gi) is adenyl cyclase, which thus helps either in the production or

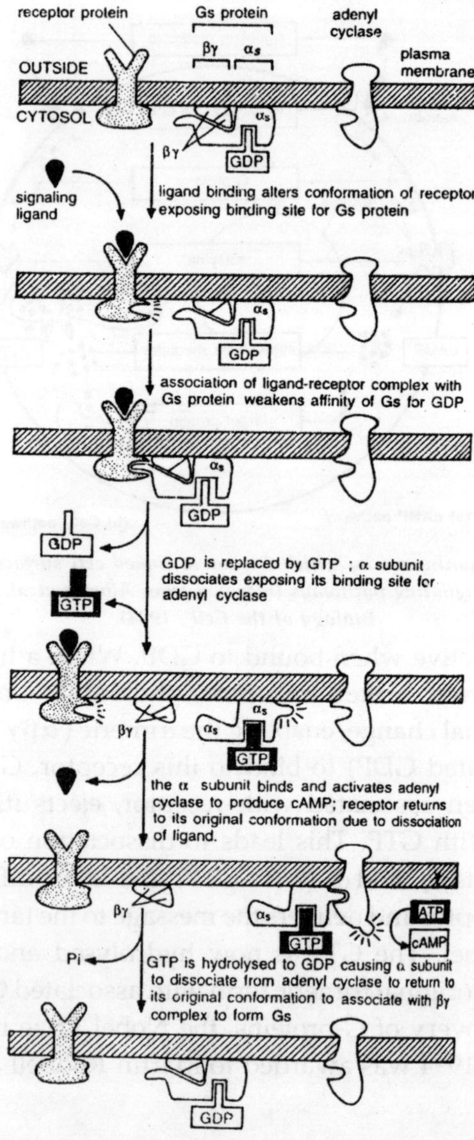

A typical trimeric (αβγ) G protein, where GTP binding releases α-subunit from βγ dimer; α-GTP subunit becomes free to act upon target protein.

Some cellular responses mediated by G-protein linked receptors.

Target tissue /organ	*Signaling molecule*	*Major response*
I. Cyclic AMP pathway		
Thyroid gland	Thyroid stimulating hormone (TSH)	Thyroid hormone synthesis & secretion
Adernal cortex	Adrenocorticotropic hormone (ACTH)	Cortisol secretion
Ovary	Luteinizing hormone (LH)	Progesterone secretion
Muscle	Adrenaline	Glycogen breakdown
Bone	Parathormone	Bone resorption
Heart	Adrenaline	Increase in heart rate and force of contraction
Liver	Glucagon	Glycogen breakdown
Kidney	Vasopressin	Water resorption
Fat	Adrenaline, ACTH, glucagon, TSH	Triglyceride breakdown
II. Inositol phospholipid pathway		
Liver	Vasopressin	Glycogen breakdown
Pancreas	Acetylcholine	Amylase secretion
Smooth muscle	Acetylcholine	Contraction
Mast cells	Antigen	Histamine secretion
Blood platelets	Thrombin	Aggregation

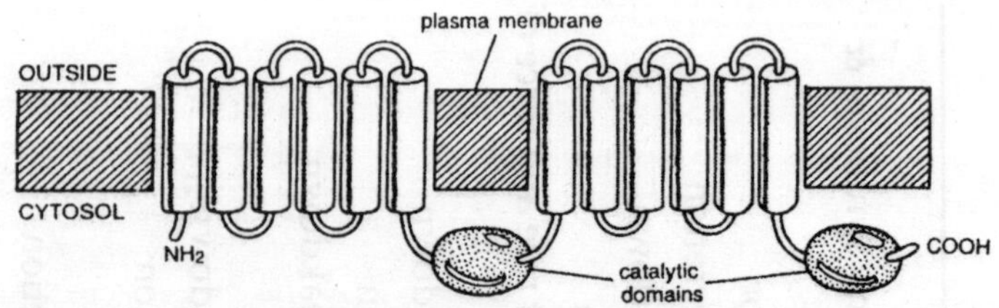

Structure of adenyl cyclase enzyme, showing two clusters of six transmembrane segments, separating two similar cytoplasmic catalytic domains, (redrawn from Alberts el al., 'Molecular Biology of the Cell', 1994).

in reduction of cyclic AMP, a second messenger which mediates several cell responses. The increase or decrease of the concentration of cyclic AMP depends on the receptor to which the ligand binds. β adrenergic receptors are used for activation and α_2-cidrenergic receptors are used for inhibiting adenyl cyclase thus influencing the concentration of cyclic AMP. The cyclic AMP exerts its influence through cyclic-AMP-dependent protein kinase, called A-kinase (PKA), which catalyzes transfer of terminal phosphate group from ATP to specific ser/thr of selected proteins. These target proteins may themselves be enzymes mediating a signal transduction pathway or a biochemical synthetic pathway (Figs.). The effects of A-kinase (PKA) may be reversed by ser/thr protein phosphatases. Another class of G protein (Gq) mediates activation or inhibition of another enzyme, phospholipase C-β which cleaves PIP_2 (PI bisphosphate; PI=phosphatidyl inositol; PI is a minor phospholipid in cell membrane) and generates two products: inositol triphosphate (IP_3) and diacylglycerol (DAG) (Figs.). Both these molecules play important roles in mediating several cell responses, through increasing or decreasing Ca^{++} concentration in the cytosol. The pathway is described as inositol phospholipid pathway (Table).

IP_3 released from PIP_2 due to phospholipase C-β diffuses rapidly through the cytosol, binds to IP3 -gated Ca^{++} release channels in ER membrane and helps in release of Ca^{++} from ER, thus increasing Ca^{++} concentration in the cytosol. Similarly diacylglycerol (the other product of hydrolysis of PIP_2) activates protein kinase C (C-kinase or PKC; so called because it is Ca^{++} -dependent; Fig.). In many cells, the activation of C-kinase increases the transcription of specific genes. This is achieved due to phosphorylation of either another kinase (e.g. MAP-kinase, thus initiating a cascade of protein kinases) or an inhibitor protein (e.g. Iκ-B) leading to release of a regulatory protein (e.g. NF- κ B; Fig.). In both cases, eventually the regulator protein (e.g. a transcription factor) stimulates the transcription of specific genes.

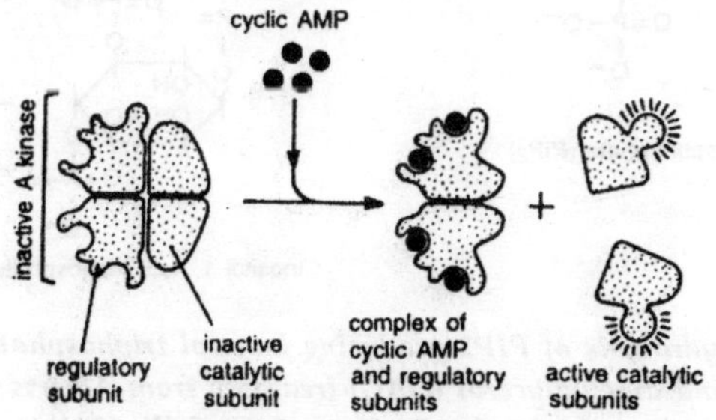

The activation of cyclic-AMP dependent protein kinase (A-kinase or PKA) due to phosphorylation of a subunit of the enzyme (redrawn from Alberts et al., 'Molecular Biology of the Cell', 1994).

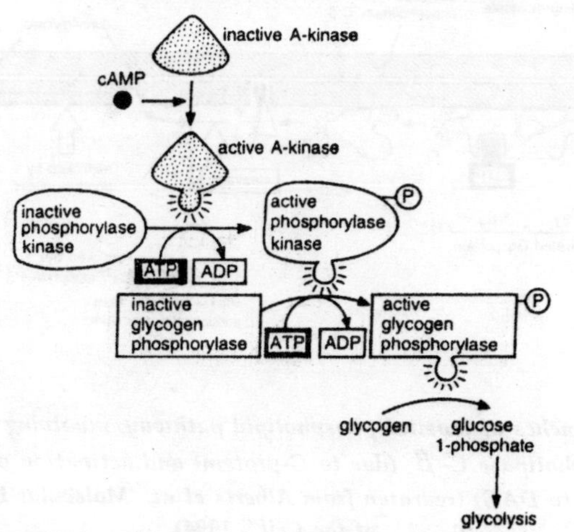

Stimulation of glycogen breakdown by cyclic AMP in skeletal muscle cells (redrawn from Alberts et al, 'Molecular Biology of the Cell', 1994).

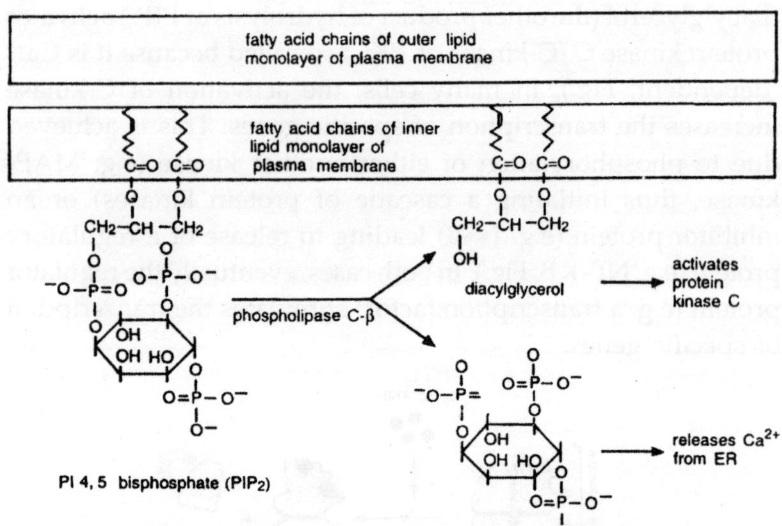

Hydrolysis of PIP2 producing inositol triphosphate (IP3) and diacylglycerol (DAG) (redrawn from Alberts et al., 'Molecular Biology of the Cell', 1994).

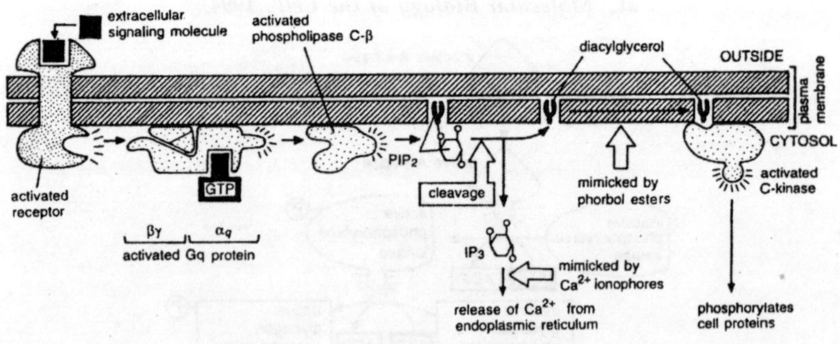

Two branches of inositol phospholipid pathway, involving activation of phospholipase C- β (due to G-protein) and activation of C-kinase (due to DAG) (redrawn from Alberts et at. 'Molecular Biology of the Cell", 1994).

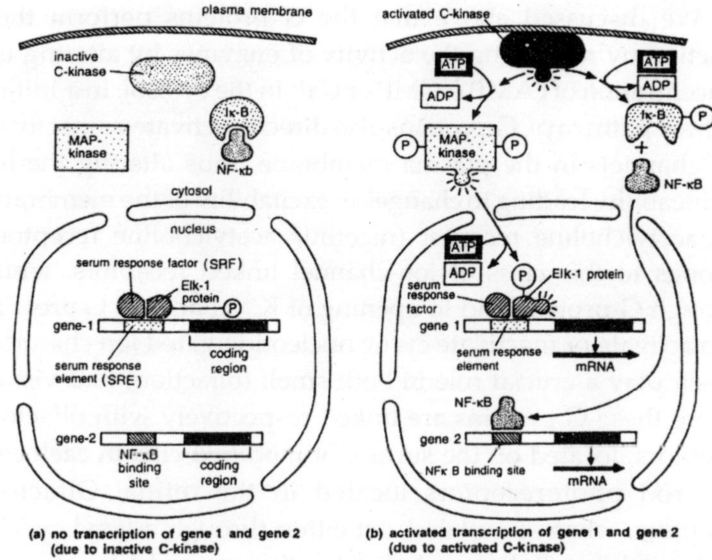

The role of C-kinase (PKC) in activation of the transcription of specific genes, through following two intracellular pathways (i) one pathway involves C-kinase phosphorylating MAP-kinase, which in its turn phosphorylates the protein Elk-1; Elk-1 along with SRF (serum response factor) activates serum response element (SRE); (ii) the other pathway involves C-kinase, phosphorylating Iκ-B, which releases NF-κB to reach the nucleus and activate transcription of specific genes (redrawn from Alberts et al., 'Molecular Biology of the Cell', 1994).

Cytosolic Ca^{++} acts as a second messenger, its concentration, influenced as above, regulates gene expression in a variety of ways. An important pathway is binding of Ca^{++} to calmodulin. The Ca^{++}/calmodulin complex has no enzyme activity and acts by altering the activity of other proteins (e.g. Ca^{++}-ATPase or CaM-kinases). The CaM kinases (Ca^{++}/calmodulin-dependent protein kinases), which mediate most of the effects of Ca^{++} in the cell, include many enzymes including the following : (i) myosin light-chain kinase (it activates smooth muscle contraction); (ii) phosphorylase kinase (it activates glycogen breakdown); (iii) CaM kinase II (Fig.; it activates secretion and resynthesis of neurotransmitter, thus helping in memory and learning).

We discussed above that the G proteins perform their function by regulating the activity of enzymes by altering the concentration of cAMP/cGMP or Ca^{++} in the cytosol. In addition to these pathways, G-proteins also directly activate or inactivate ion channels in the plasma membrane, thus altering the ion permeability leading to changes in excitability of the membrane. An acetylcholine receptor (nicotinic acetylcholine receptors) belongs to this class of ion-channel linked receptors, which through G-protein lead to opening of K^{++} channels. G proteins also activate or inactivate cyclic nucleotide-gated ion channels, which play a crucial role in both smell (olfaction) and vision, where these G proteins are linked respectively with olfactory receptors, located on the surface of modified cilia in each cell, and rod photoreceptors located in the retina. Olfactory receptors, when stimulated, act either through adenyl cyclase or through inositol phospholipid pathways and IP_3-gated ion channels. Similarly, rod photoreceptors are coupled with G-protein (Gt or transducin), which activates guanyl cyclase or cyclic GMP phosphodiesterase. For the discovery of G-proteins, the Nobel Prize in Medicine for the year 1994 was awarded to Martin Rodbell and Arthur Gilman.

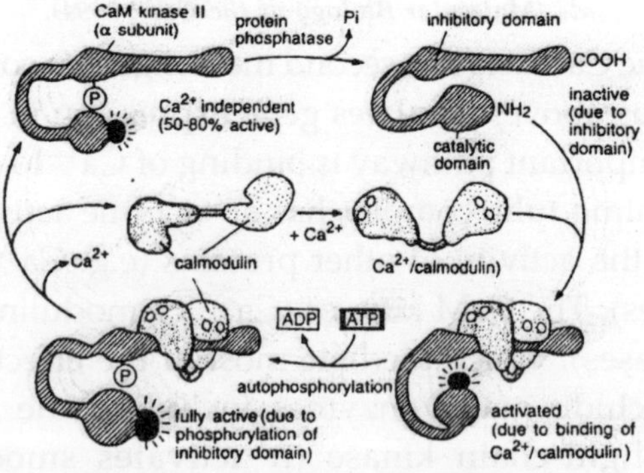

The activation of CaM kinase II; it consists of 12 subunits of four kinds (α, β, γ, δ), but only one (α), expressed in the brain, is shown in this figure; Ca2+/calmodulin helps in autophosphorylation of the enzyme, thus rendering it Ca^{2+} independent (redrawn from Alberts et al., 'Molecular Biology of the Cell, 1994).

Enzyme-linked Cell Surface Receptors : Enzyme linked receptors are also transmembrane proteins like the G-protein linked receptors. Their cytosolic domains either have an enzyme activity (as in several tyrosine and serine/threonine kinases) or are associated directly with an enzyme (as in JAK-STAT pathway, discussed later in some detail in this Chapter). There are five classes of enzyme linked receptors ; (i) receptor guanyl cyclases (catalyze production of cyclic GMP); (ii) receptor tyrosine kinases (catalyze phosphorylation of specific intracellular proteins); (iii) tyrosine-kinase associated receptors (associate with proteins having tyrosine kinase activity); (iv) receptor tyrosine phosphatases (remove phosphate group from tyrosine) and (v) receptor serine/threonine kinases (catalyze phosphorylation of specific serine/ theronine residues of specific intracellular proteins). Some of the details of these five types of receptors, including suitable examples are summarized in Table. In all these cases, ligand binding usually induces the receptors to dimerize, which activates the kinase domain of the receptor or an associated non-receptor protein kinase. When the receptor is activated, two molecules of a tyrosine kinase in a dimer, cross-phosphorylate themselves on multiple tyrosine residues (only dimerized receptors are active). The tyrosine phosphorylated proteins are the binding sites of intracellular signaling proteins, which, via their SH_2 domains, bind to specific phosphotyrosine residues. In this way, a multiprotein signaling complex is activated from which the signal spreads to the cell interior.

Cell surface receptors in cholesterol metabolism and drug production (1985, '88 Nobel Prizes). During last more than two decades, a study of regulation of cholesterol metabolism was conducted by S. Brown and J.L. Goldstein (at University of Texas, U.S.A.), both of them sharing the Nobel Prize for Physiology or Medicine in 1985. Their study mainly involved the study of cell surface receptor for low density Iipoprotein (LDL), which is a major transporter of blood cholesterol from blood to the interior of fibroblast and liver cells.

Five different types of enzyme linked cell surface receptors and their characteristics.

Ligand	*Receptor*	*Catalytic site*	*Response on ligand binding*
I. Receptor guanyl cyclase			
1. Atrial natriuretic peptides or ANPs (a peptide hormone secreted in atrium of heart, when BP rises)	ANP receptors in kidney cells & smooth muscle cells of blood vessels	Guanyl cyclase site in cytosolic domain of receptor protein (no G-protein is involved)	Cyclic GMP is produced and activates G-kinase which phosphorylates ser/thr in specific proteins
II. Receptor tyrosine kinases			
2. Epidermal growth factor or EGF (a small protein, 53 amino acids long)	EGF receptor (single pass transmembrane proteins; 1200 amino acids)	One tyrosine kinase cytosolic domain	Autophosphorylation followed by phosphorylation of a cascade of intracellular signaling proteins having SH2 and SH3 (SH stands for Src homology) domains (e.g. GTPase- activating protein or GAP, phospholipase C)
3. Nerve growth factor or NGF	NGF receptor	Two tyrosine kinase cytosolic domains	Phosphorylation of receptor
4. Insulin and insulin like growth factor (IGF-1)	Insulin & IGF-1 receptors (a tetramer)	Two tyrosine kinase cytosolic domains	Allosteric interaction of autophosphorylation of receptor

Contd..

Ligand	*Receptor*	*Catalytic site*	*Response on ligand binding*
5. Platelet derived growth factor or PDGF and microphage colony stimulating factors or M-CSF	PDGF receptor & M-CSF receptor	One tyrosine kinase cytoslic domain interrupted by a 'kinase insert region'	Autophosphorylation of activated receptor
6. Fibroblast growth factors or FGFs	FGF receptor	"	"
7. Vascular endothelid growth factor (VEGF)	VEGF receptor	"	"
III.Tyrosine kinase associated receptors			
8. Cytokines (e.g. interleukins like IL-2)	Cytokine receptors (e.g. IL-2 receptor)	Janus family of non-receptor tyrosine kinases (Jakl, Jak2, Tyk2)	Assembly of functional receptor, which activates a tyrosine kinase
IV.Protein tyrosine phosphatases			
9. Foreign antigens	CD45 protein on the surface of WBCs	Tyrosine phosphatase cytosolic domain	Activation of T and B lymphocytes
V.Receptor serine /threonine kinases			
TGF-β superfamily (e.g.TGF-β1 to β5, activins, bone morphogenetic proteins)	TGF-β receptor (single pass transmembrane protein)	Serine-threonine kinase cytosolic domain	Activation of kinase domain

Table. In all these case, ligand binding usually induces the receptors to dimerize, which activates the kinase domain of the receptor or an associated non-receptor protein kinase. When the receptor is activated, two molecules of a tyrosine kinase in a dimer, cross-phosphorylate themselves on multiple tyrosine residues (only dimerized receptors are active).

The tyrosine phosphorylated proteins are the binding sites of intracellular signalling proteins, which via their SH_2 domains, bind to specific phosphotyrosine residues. In this way, a multiprotein signalling complex is activated from which the signal spread to the cell interior.

Cell Surface Receptors in Cholesterol Metabolism and Drug Production (1985, '88 Nobel Prize). During last more than two decades, a study of regulation of cholesterol metabolism was conducted by S. Brown and J.L. Goldstein (at University of Texas, U.S.A.) both of them sharing the Nobel Prize for Physiology or Medicine in 1985. Their study mainly involved the study of cell surface receptor for low density lipoprotein (LDL), which is a major transporter of blood cholesterol from blood to the interior of fibroblast and liver cells.

The receptor molecule forms a complex with LDL and enters the cell through pinocytosis. The receptor is recycled and LDL is degraded so that cholesterol can be utilized in the cell. When a patient has a defective gene for receptor, it leads to high blood cholesterol and LDL.

This leads to the development of familial hypercholesterol-anemia or coronary atherosclerosis (before the age of 20 in a homozygous defective). Brown and Goldstein also synthesized the gene for enzyme HMGCoA reductase, which is responsible for cholesterol synthesis and is under the control of a feedback inhibition. These researches may eventually lead to prevention and treatment of atherosclerosis.

Signal Transduction Pathways

Signal Transduction Using 'Second Messengers : In animals as well as in plants, the cellular machinery for protein synthesis responds to external stimuli. These external stimuli may include a whole range of diverse factors including the following : (i) the presence of hormones, growth factors and other signalling molecules (e.g. in animals); (ii) the presence of pheroinones secreted by cells of two different mating types (i.e. in yeast);

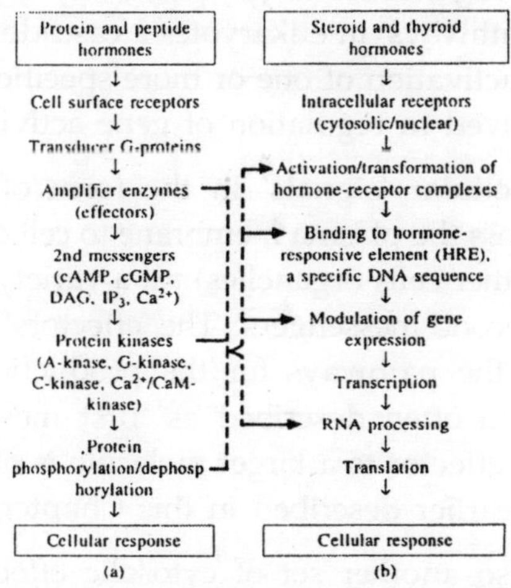

Different steps involved in "signal transduction pathways' for regulation of gene expression due to action of (a) protein/peptide hormones; (b) steroid and thyroid hormones (also note cross-linking between two pathways).

(iii) presence of light (different wave lengths), gravity, minerals, water, gases and specific soil structure (e.g. in plants) and (iv) cell-cell contact. In some of these cases, cell surface receptor molecules are present and transmit the signal to the interior of the cell (for water-soluble signals like proteins, peptides and neurotransmitters). In other cases, substances like hormones interact with intracellular receptors, which transmit the signal further to the target site (for water insoluble lipophilic signals like steroids, thyroid hormones, retinoic acid, etc.). In all these cases (including prokaryotes), the external stimulus needs to be transduced and integrated with internal signals, which follow a cascade of events called- 'signal transduction pathway '- leading to the response, which is a specific gene expression in many cases (Fig.).

Signal transduction pathways precede most of the mechanisms for regulation of gene expression discussed earlier. In prokaryotes, often the precursor molecule (e.g. lactose) or the end product (e.g. an amino acid) of a biosynthetic pathway takes part in regulation of gene activity. But in 'signal transduction pathways' in eukaryotes a cascade of molecules, leading to the activation of one or more specific transcription factors, is involved in regulation of gene activity.

The extracellular 'signals' in the form of ligands, are transduced across the plasma membrane to cell compartments (nucleus and other cells organelles) *via* a variety of molecules described as 'second messengers'. The 'effectors' i.e. molecules which initiate the pathways for the production of 'second messengers', are often described as 'first messengers'. The binding of this effector to a target molecule is often mediated by G-proteins earlier described in this Chapter.

There is also another set of cytosolic effectors, through which second messengers' produce a wide variety of cellular responses. Although there are millions of molecules which send signals into the cell interior, only half a dozen or so 'second messengers' have so far been identified (e.g. cAMP, cGMP, DAG, IP_3, Ca^{2+}, etc.). Actually, signal transduction pathways are also known, which do not involve any 'second messenger' molecules (e.g. response to interferon a, insulin, steroid and thyroid hormones, etc.). There are also extracellular proteins (e.g. lactoferrin) which act as their own messengers for signal transduction.

Cyclic AMP and Ca^{++} Mediated Signal Transduction : As discussed earlier several cell responses are mediated by alterations in the concentration of simple chemicals like cyclic

AMP (cAMP) and calcium ions (Ca^{++}) within the cell. This is achieved by activation of cell surface receptors due to either the attachment of regulatory molecules (e.g. hormones) or due to other physical factors.

The activation of receptor leads to activation of a G-protein, which in its turn activates the amplifier enzyme, adenylate cyclase (an effector). This enzyme converts ATP into cAMP, leading to an increased concentration (inside the cell) of cyclic AMP, Ca^{++} or whatever the internal mediator is. Since cyclic AMP conveys the message in the interior of the cell, it was described as a 'second messenger'. In most cases, cAMP activates a protein kinase A (PKA), which phosphorylates several specific enzymes (e.g. lipase for mobilization of fat; phosphorylase for enzymatic breakdown of glycogen in the liver). In addition to its role in activation of several cellular enzymes, cAMP often causes an elevation of cytosolic level of Ca^{++}, which is another second messenger. Epinephrine (adrenaline hormone) also acts through cAMP to modulate the intracellular level of Ca^{++} in muscles and nerve tissues.

Lipids (DAG, IP_3) and Ca^{++} Mediated Signal Transduction : Some external signals have been found to stimulate the turnover of a membrane lipid and change the intracellular concentration of Ca^{++}. In this connection, the metabolism of phosphatidylinositides (Pis) has been shown to play an important role in the regulation of cell growth. It was also shown that increased PI metabolism is a reflection of signal transduction across the cell membrane. In 1980's, the activation of receptors could be linked to the rapid hydrolysis of Pis and mobilization of Ca^{++} in many tissues in response to a variety of external stimuli. As discussed earlier in this chapter hydrolysis of a PI (phosphatidylinositol 4,5 biphosphate = PIP_2) utilizes an amplifier enzyme, the Pi-specific phospholipase Cβ (PLC) and produces various messenger molecules. The following three 'lipid second messengers' have been documented : (i) 1,2-diacylglycerol (DAG); (ii) inositol 1,4, 5-triphosphate (Ins 1, 4, 5, – P_3 = IP_3) and (iii) archidonic acid. The hydrolysis of PIP_2, upon binding of extracellular ligands to cell surface receptors is also coupled to G-protein activation. Activated G-protein modulates PLC, which in turn hydrolyses PIP_2 into DAG and IP_3. Although, the signal transduction utilizing these lipid messengers has been shown to be complex, it is known that DAG serves as a co-factor and together with Ca^{++} activates protein kinase C (PKC). During this process, DAG itself is metabolized to archidonic acid, a precursor of other mediators. (PKC plays a central role in the transduction of extracellular signals into a cellular response). On the other hand, IP_3 promotes the release of Ca^{++} into the cytoplasm through binding to a specific receptor on the surface of endoplasmic reticulum (ER) and/or other intracellular vesicles that store Ca^{++}.

This interaction induces a conformational change in proteins specific to Ca^{++} channel and allows Ca^{++} ions to exit from intracellular storage sites into the cytoplasm. This calcium influx following stimulation of the PI signal pathway is a widespread and physiologically most important phenomenon. Ca^{++} in the cytoplasm has several roles to play in the cellular regulation, including activation of PKC by DAG. In view of the role of IP_3 as a 'second messenger' in operating Ca^{++} channels as above, these Ca^{++} channels have been described, in the past, as SMOCs (second messenger operated calcium channels). However, since the role of 'second messengers' is not fully established, the usage of the term SMOC has been questioned.

Ras-mediated Signal Transduction Pathways : Receptor tyrosine kinases and G-protein linked receptors, both initiate intracellular signaling cascades in which Ras proteins (product of proto-oncogene *c-ras*, related with cancer belonging to Ras superfamily of monomeric GTPases, play an important role. Like G-proteins and other GTPases, Ras proteins are active, when bound to GTP and inactive when bound is GDP. Two classes of signaling proteins, that regulate Ras activity include the following : (i) GTPase-activating proteins (GAPs) increase the rate of hydrolysis of bound GTP by Ras, thereby inactivating Ras. (ii) Guanine nucleotide releasing proteins (GNRPs) promote the release of GDP and uptake of GTP, thus restoring the activity of Ras. Therefore receptor tyrosine kinases activate Ras either by inhibiting a GAP or by activating a GNRP. When ras proteins are active, they initiate a cascade of ser/thr phosphorylations, that converge on MAP-kinases, which help relay the signal to the nucleus. Many of these intracellular signaling proteins, when inappropriately activated, can cause cancer.

Ras Mediated Signal Transduction for Cell Division. It has been known for some years, that signals received at the cell surface are transmitted via activated tyrosine kinase (associated with the receptor) in the cell membrane, through serine/threonine kinase in the cytoplasm, to the transcription factor in the nucleus (among other targets). One such signal transduction pathway, discovered fully in 1992-1993, involved a flow of signal via Ras to Raf-1 kinase (product of another oncogene, raf-1), thence to MSP kinases and pp^{90rsk} and finally on to nuclear proto-oncogene products like c-Fos, c-Jun, etc. (MAP = mitogen activated protein; pp^{90rsk} = ribosomal S6 kinase).

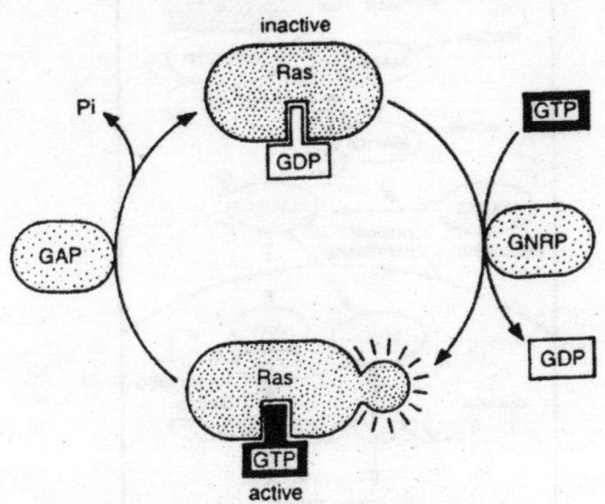

Regulation of Ras activity by GAPs (GTPase activating proteins) or by GNRPs (guanine nucleotide releasing proteins).

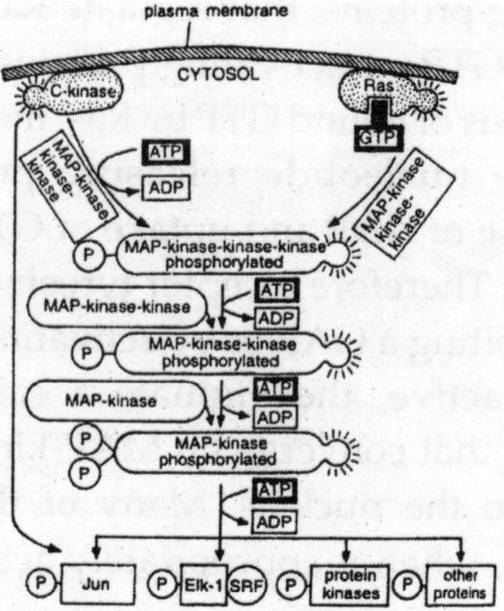

Serine-threonine phosphorylation cascade activated by Ras and C-kinase and involving a series of MAP kinases.

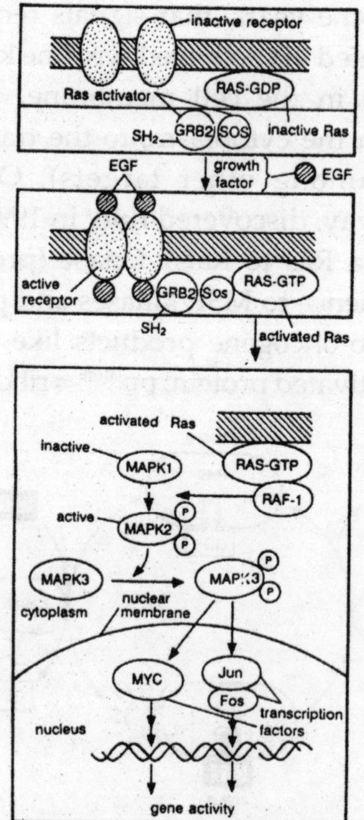

Different steps involved in Ras mediated signal transaction pathway for inducing cell division.

Following steps, shown in Figure are involved in EGF-induced Ras signal transduction pathway, leading to cell division: (i) EGF binds to receptor to form activated EGF-receptor complex, which is an active tyrosine kinase and phosphorylates itself; (ii) phosphorylation of receptor, enables its internal segment (cytosolic) to bind to GRB2 protein in the

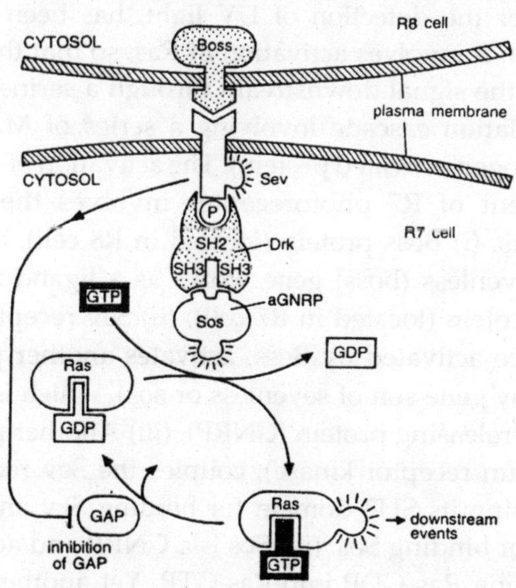

Regulation of Ras activity in early cell signaling leading to R7 development; note the presence of Boss protein (in R8 cell), which activates sev receptor protein in R7 precursor cell, leading to the activation of Ras.

GRB2-mSos complex (GRB2 and mSos are two intermediary proteins, which remain linked together in unstimulated cells); (iii) mSos of the GRB2-mSos complex now attaches to Ras, which is anchored to the inside of the cell membrane; mSos activates Ras by changing it from inactive Ras-GDP to active Ras-GTP; (iv) Raf-1 now binds to activated Ras-GTP and gets activated; (v) Activated Raf-1 now activates MAP kinases (MAPK) forming a 'kinase cascade', (each MAPK phosphorylates, and thus activates the next member of the series); (vi) The last MAP kinase phosphorylates several cell proteins, including some transcription factors (e.g. Myc, Fos,

Jun, etc.); (vii) Phosphorylation of transcription factors leads to expression of specific genes for cell division.

Ras mediated Development of R7 Photoreceptor in Drosophila: A compound eye in Drosophila consists of 800 ommitidia, each consisting of 8 photoreceptor cells (designated as R1-R8) and 12 accessory cells. The development of R7, required for the detection of UV light, has been studied in some detail. It involves activation of Ras, so that the activated Ras relays the signal downstream through a serine/threonine phosphorylation cascade involving a series of MAP kinases (MAP=mitogen activated protein). The activation of Ras, during development of R7 photoreceptor involves the following components. (i) Boss protein (located in R8 cell), encoded by bride-of-sevenless (boss) gene works as a ligand for the Sev receptor protein (located in R7 cell); (ii) Sev receptor tyrosine kinase, once activated by Boss, activates another protein Sos (encoded by gene son of sevenless or sos), which is a guanine nucleotide releasing protein (GNRP). (iii) Another protein Drk (downstream receptor kinase), couples the Sev receptor with the Sos, using its SH2 domain for binding Sev and two SH3 domains for binding Sos. (iv) Sos is a GNRP and activates Ras by converting Ras-GDP into Ras-GTP. Yet another protein, a GTPase activating protein (GAP), facilitates hydrolysis of Ras bound GTP, thus converting active Ras-GTP into inactive Ras-GDP. If the gene for GAP mutates, R7 develops even in the absence of Sev, due to the availability of hyperactive Ras.

JAK-STAT Signal Transduction Pathway (Mediated by Cytokine Receptors) : In 1994, a study of the IFN-α and IFN-γ induced transcriptional activation, led to the discovery of a new signal transduction pathway called JAK-STAT pathway (JAK stands for Janus kinases and STAT refers to cytoplasmic transcription factors called STATs; STAT stands for Signal Transducers and Activators of Transcription). Tyrosine kinases of JAK family remain closely associated with the components of transmembrane receptors and are activated after ligand-receptor attachment (ligands are polypeptides like IFN-α & IFN-γ, that are recognized by ligand binding domain of a receptor protein found in plasma membrane. The receptor consists of two domains, the ligand binding domain α and a signal transducing chain β). This leads to phosphorylation of one or more of the transcription factors called STATs. The STATs perform the following two roles: (i) they work as signal transducers by acting as substrates of the JAKs, and undergo phosphorylation leading to their own activation;

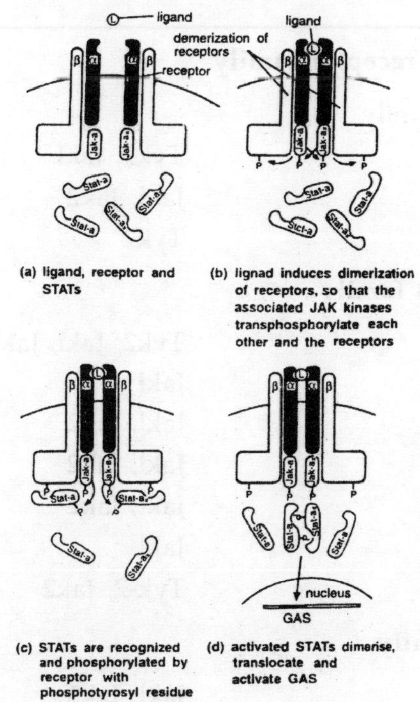

An outline of JAK-STAT signal transduction, explaining the roles of Janus kinases (JAKs) and STATs.

(ii) after phosphorylation, homo/ heterodimerisation and nuclear translocation, STATs act as transcriptional activators and activate specific genes. This JAK-STAT pathway was though discovered due to a study of IFN-α and IFN-γ induced activation of transcription, many other JAK and STAT proteins were later found to follow this pathway.

Receptors and Kinases of JAK-STAT pathway : A large number of receptors, each representing a transmembrane protein are known. Following two families of these receptors are involved in JAK-STAT pathway:

Activation of JAK kinases by various ligands

Ligands	*Receptor*	*STATs activated*
I. Cytokine receptor family		
IFN family		
IFN -	Tyk2, Jakl	1,2,3
IFN	Jakl, Jak2	1
IL-10	Tyk2	1,3
gp 130 family		
IL-6	Tyk2, Jakl, Jak2	1,3
IL-11	Jakl	1,3
OnM	Jakl, Jak2	1,3
LIF	Jakl, Jak2	1,3
CNTF	Jakl, Jak2	1,3
G-CSF	Jakl	1,3
IL-12	Tyk-2, Jak2	3,4
-C family		
IL-2	Jakl, Jak3	1,3,5
IL-4	Jakl, Jak3	6
IL-7	Jakl, Jak3	5
1L-9	Jakl, Jak3	5
IL-13	Jakl	
IL-15	Jakl, Jak3	5

Contd...

Ligands	*Receptor*	*STATs activated*
gpl40 family		
IL-3	Jak2	5
IL-5	Jak2	5
GM-CSF	Jak2	5
Growth hormone family		
EPO	Jak2	5 ?
GH	Jak2	1,3,5
PRL	Jak2	1,5
II. Receptor tyrosine kinases		
EGF	Jakl, Jak2	1,3
PDGF	Jakl, Jak2	1,3
CSF-1	Jakl, Jak2	1,3

(i) Cytokine receptor family; (ii) Tyrosine kinase receptor family. The ligands that interact with these two classes of receptors are listed in above Table, where the JAK kinases and STATs activated by each of these ligands are also shown. There are four JAK kinases, called Tykl, Jakl, Jak2, Jak3, all of them lacking SH2 and SH3 domains. These kinases remain inactive, but remain associated with receptors, so that they are rapidly activated by a ligand stimulated phosphorylation on a tyrosine. The activated JAK kinases mediate activation of STAT proteins, leading to transcription activation of specific genes.

The following steps involved in this signal transduction pathway are shown in Figure. (i) ligand promotes dimerization of signal transducing component (p chain) of receptor; (ii) two molecules of the JAK kinase, which remain associated with specific domains of the receptor are drawn in apposition and transphosphorylate each other; (iii) the kinases then phosphorylate a tyrosine in a distal region of receptor P-chain, (iv) active receptor draws STAT molecules through interaction with STAT SH2 domain, (v) STAT molecules are phosphorylated by JAK kinase, making them competent to dimerize, travel to the nucleus and bind to specific response elements in DNA.

Steroid Hormones Induced Signal Transduction. It has been demonstrated that sex steroid hormones, namely estrogen and progesterone can regulate gene activity. Glucocorticoids, another set of steroids, found endogenously within the cells and used for treatment of leukaemia, asthma and arthritis also regulate gene expression. These glucocorticoids have a multitude of effects in various tissues. Classical example of a sex steroid inducible protein is chick ovalbumin, and two of the glucocorticoid-inducible proteins are the enzymes tyrosine aminotransferase (TAT) and tryptophan oxydase. The glucocorticoids induce synthesis of these enzymes through receptor molecules. The steroid-receptor complex interacts each with a specific DNA sequence, called steroid receptor element (SRE) or glucocorticoid receptor element (GRE) or hormone receptor element (HRE). Through recombinant DNA technology, gene constructs were prepared by fusing different lengths of DNA sequences flanking the 5' end of TAT gene with the marker gene for CAT (chloramphenicol acetyltransferase). These cloned gene constructs were introduced into animal cells and though glucocorticoid induced expression of CAT, steroid receptor elements (SRE) were identified.

Steroid inducible synthesis of protein, which involves the following six steps, is illustrated in Figure.

(i) Uptake of steroid hormone (S) and binding to receptor molecule in the cytoplasm (Rc) to form S-Rc complex. (Receptors for gluco-corticoids are predominantly found in the cytoplasm, but those for sex steroids are found in the nucleus),

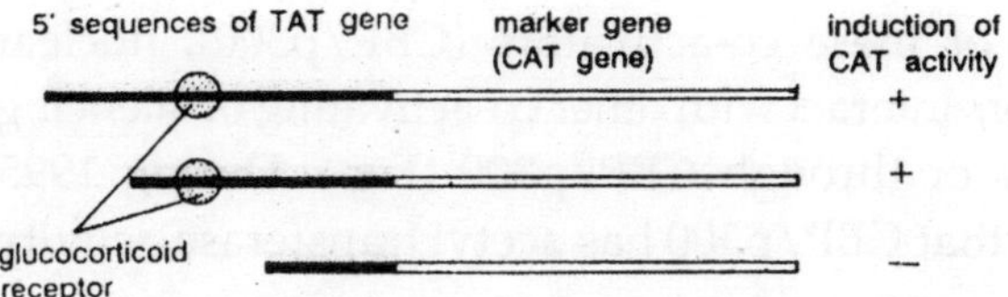

Use of deletions through recombinant DNA technology for identification of response elements.

S S S S S S → Rc SRc → SRn
Rc
Rc
Ac
SRn
Ac = Acceptor site
R = Receptor
S = Steroid
SRc = steroid-receptor complex in the cytoplasm
SRn = steroid-receptor complex in the nucleus

Role of steroid hormones in regulation of gene activity in eukaryotes.

(ii) Transport of S-Rc complex to nucleus (wherever cytosolic receptor is involved), where it is now called S-Rn to specify its occurrence in the nucleus,

(iii) Binding of S-Rn complex to specific acceptor site SRE on genome,

(iv) Activation of transcriptional apparatus in presence of specific non-histone proteins to synthesize specific mRNA.

(v) Transport of specific RNA to cytoplasm,

(vi) Synthesis of protein (like ovalbumin) and occurrence of specific steroid mediated functional response.

The nuclear hormone receptor in the above pathway needs a co-activator to initiate transcription with the help of RNA polymerase II. This co-activator is CBP or related p300 (CBP=CREB binding protein; CREB=cyclic AMP response element binding regulatory protein). CBP/p300 is a large protein, that interacts with different DNA-binding transcription factors (inducing nuclear hormone receptor) and also with (he components of the basal transcription machinery. In addition to one of these co-activators (CBP/p300), nuclear hormone receptors interact with other co-activating or silencing mediators directly or through CBP/p300 (Fig.). During 1995-97 it was shown that CBP/p300 has acetyl transferase activity and helps

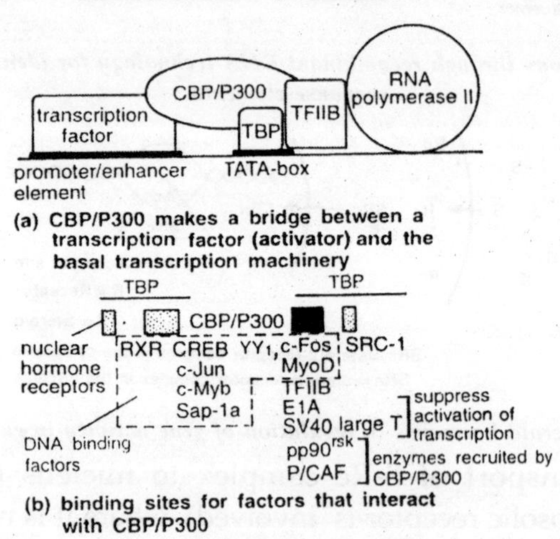

The role of CBP/p300 (a protein complex), as a co-activator, in steroid inducible synthesis of proteins.

in acetylation of histone proteins, which makes chromatin's DNA more accessible to the transcriptional machinery. There are also deacetylases, (e.g. Rpd3 = reduced phosphate dependency), which remove acetyl groups from histone proteins and stabilize the repressed state of chromatin thus working as corepressors (Fig.).

Interferon Stimulated Gene Expression (without a Second Messenger) : Interferons were once believed to have the ability to fight cancer and, therefore, received major attention of molecular biologists, even though its early clinical promise was not fulfilled. However, recent studies with interferon gave information about how cells really respond to interferon signals. In the year, 1992, several teams of researchers demonstrated that when interferon contacts its receptor on the cell surface, it immediately activates a transcription factor which moves to the nucleus, where it turns on its own particular set of genes. It was experimentally demonstrated that no 'second messengers' are involved in this signal transduction pathway.

In mid-1980's, the genes that respond to interferon were shown to carry a common regulator sequence called ISRE (interferon stimulated response element). In late 1980's, the transcription factor that binds to ISRE was also discovered.

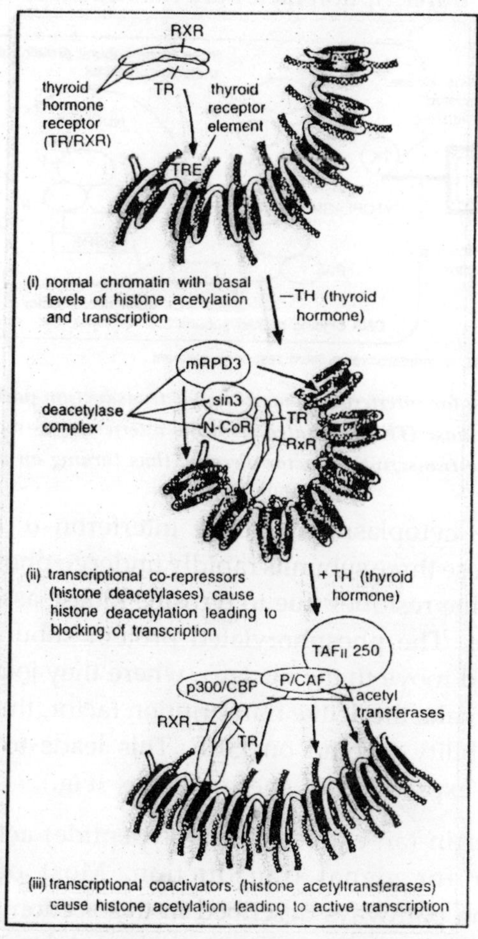

Acetylation and deacetylation of histone proteins and their role in the regulation of transcription.

This transcription factor consists of four protein subunits including three large subunits, and a small subunit. The fourth small subunit actually makes contact with the ISRE. It was also shown that the three large subunits of the transcription factor, each contains an SH2 domain, a sequence commonly found in proteins that interact with tyrosine kinase. In cells, which have not been stimulated by interferon-α , the three large subunits of the above transcription factor are not linked and are located

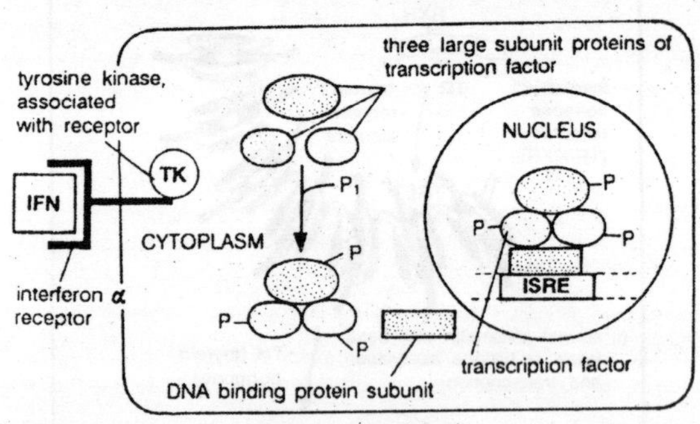

A direct route for interferon induced 'signal transduction pathways' where the tyrosine kinase (TK) associated with the interferon α -receptor complex activates the transcription factor directly, thus turning on the interferon responsive genes.

only in the cytoplasm. But once interferon-α binds to its receptor, these three subunits rapidly undergo phosphorylation at the tyrosine residues due to tyrosine kinase associated with the receptor. The phosphorylated protein subunits associate together and move to the nucleus, where they join the smaller subunit to make the active transcription factor, thus conferring on it the ability to bind on ISRE. This leads to interferon-a stimulated expression of specific genes (Fig.).

Lactoferrin (an extracellular polypeptide) acts as its own messenger for signal transduction. Most of the signal transduction pathways described in this section are complex, involving as many as seven or more polypeptides, which deliver the signal from the membrane receptor to the cell nucleus.

However, in Jak-STAT pathway and in the pathways utilized by steroid hormones, fewer intermediaries are used to relay the message. In 1995, an extracellular molecule, lactoferrin (a secretory protein) was found to carry the message all by itself from cell surface to the cell nucleus where, in its dimeric or tetrameric form, it binds to the sequence 5'-GGCACTTGC-3'. No cell surface receptor or second messenger is involved in

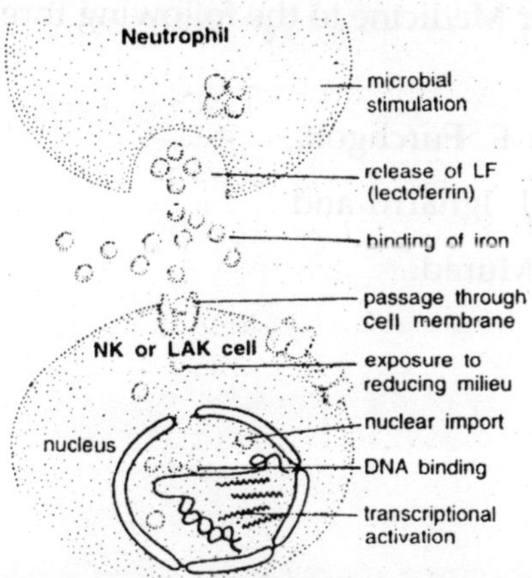

Lactoferrin induced activation of transcription in which lactoferrin is the sole molecule carrying message from cell surface to the cell nucleus.

this pathway. Lactoferrin, therefore, is a secretory, membrane permeable transcription factor, which is also a DNA binding protein. Lactoferrin is stored in secretory granule of neutrophils, until a pathogen stimulates its release. Although a part of the released lactoferrin is used to suppress microbial growth, most of it crosses the plasma membrane or endosomal membrane of natural and' lymphokine-activated killer cells (NK and LAK cells), and there activate appropriate genes in the nucleus (Fig.). Therefore, this pathway followed by lectoferrin (perhaps also by Tat in HIV and Tax in HTLV-1), is highly effective and rapid signaling mechanism between two cells in an emergency (invasion by microbes).

Nitric Oxide as a Signaling Molecule (1998 Nobel Prize): Nitric oxide (NO) is another signaling molecule, which neither needs vesicular transport (it diffuses through plasma membrane) for secretion, nor does it need any receptor at the target cell. It plays a vital role in vasodilation (dilation of blood vessels; deficiency of NOcauses angina), in the control of viral/bacterial infection and in the neurotransmission (including penile erection). The work on the role of NO as a signaling molecule was recognized in 1998 by the award of Nobel Prize for Medicine to the following three scientists from USA :

(i) Robert F. Furchgott,

(ii) Louis J. Ignarro and

(iii) Ferid Murad.

8

Protein Synthesis

The synthesis of protein from mRNA involves translation of the language of nucleic acids (available in the form of mRNA) into language of proteins. This is much more complex than transcription and involves initiation, elongation and termination of polypeptide chains. The steps involved in translation, are diagrammatically represented in Figure, for prokaryotes and in Figure, for eukaryotes, and will be discussed in this chapter. However, for initiation and elongation of a polypeptide, the formation of aminoacyl transfer RNAs (aa-tRNAs) is a prerequisite, which will be described first, before describing the steps involved in translation.

Aminoacyl tRNA's Formation

Amino Acid Activation : This reaction is brought about by the binding of an amino acid with ATP and is mediated by specific activating enzymes known as amino acyl tRNA synthetases or aaRs (amino acid is abbreviated as aa). As a result of this reaction between amino acid and" adenosine triphosphate, mediated by specific enzyme, a complex (amino acyl-AMP- enzyme complex) is formed. It should be realized that amino acyl-RNA synthetases are specific with respect to amino acids. Therefore, for different amino acids, different enzymes would be required. Moreover, a particular enzyme has a double specificity in so far as it can recognize its own amino acid on the one hand and its own tRNA on the other hand. This latter specificity allows the second step outlined in the following part of this book.

$$aa_1 + ATP \xrightarrow{(Enzl)} (aa_1\text{–}AMP)\ Enz_1 + PP$$

Amino Acid Transfer to tRNA : The amino acyl-AMP-enzyme complex, formed during the step outlined above, reacts with a particular tRNA and transfers the amino acid to the tRNA. This reaction is accompanied with the liberation of

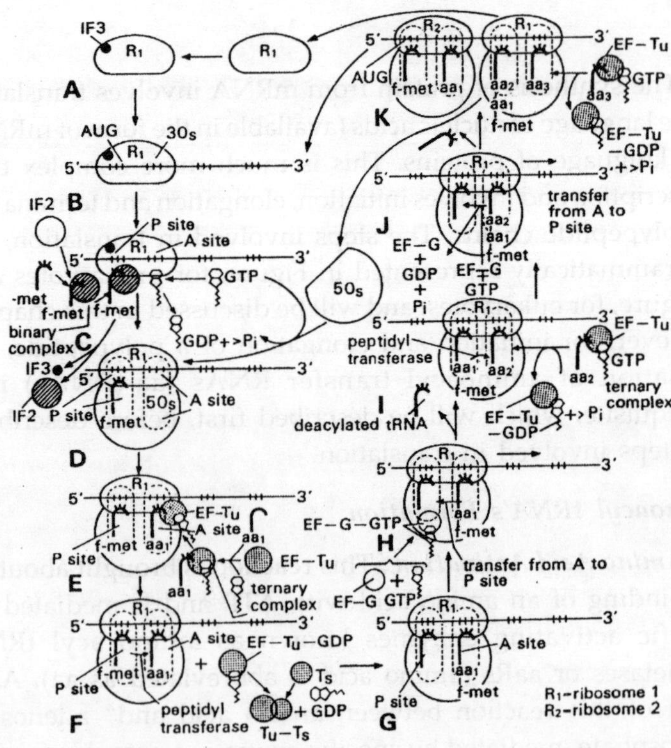

Different steps (A–K) in the translation of mRNA molecule in bacterial systems.

AMP and the enzyme. It should be noticed in this reaction that with Enz, and $tRNA_1$, 1 is used as suffix (subscript) to emphasize the specificity, which exists between them.

A particular amino acid would require a particular enzyme and a particular species of tRNA. This would mean that for 20 amino acids, at least 20 different enzymes and also at least 20 different tRNA species would be required (while there are cases where more than one kind of tRNA may be available for the same amino acid, no cases are known, where the same tRNA can recognize two or more different amino acids).

$$(aa_1 - AMP)\ Enz_1 + tRNA_1 \rightarrow aa_1 - tRNA_1 + AMP + Enz_1$$

Although, the above reaction is generally catalyzed by a protein (aaR enzyme), in recent years (1995), it was possible to use a specific selected RNA molecule, to catalyze the above reaction. It is an example, where RNA molecule can perform the same catalytic function, which is generally performed by a protein, i.e. aminoacyl-transfer RNA synthetases (aaRs). This gives support to the hypothesis of ancestral RNA world, where RNA performed the functions of nucleic acids (genetic information) and proteins (catalysis).

Initiation of Polypeptide

The initiation of polypeptide chain is always brought about by the amino acid methionine, which is regularly coded by the codon AUG, but rarely also by GUG as the initiating codon. Very rarely initiation may also be directed by UUG (~ 1%) or AUU (only one case reported). In E. coli, there are two tRNAs for methionine, $tRNA_m^{met}$ (non-formylatable) and $tRNA_f^{met}$ (formylatable). The methionine molecule carried by $tRNA_m^{met}$ cannot be formylated, while that carried by $tRNA_f^{met}$ can be formylated. In prokaryotes, formylation of initiating amino acid methionine is essential requirement, so that $tRNA_f^{met}$ is meant for depositing methionine as the first amino acid,

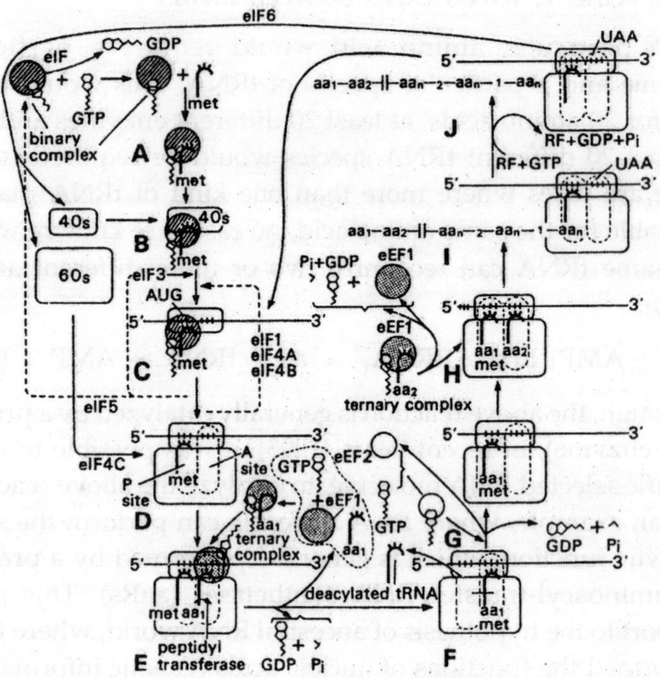

Different steps (A–J) in the translation of mRNA molecule in eukaryotes (reticulocytes)

while $tRNA_m^{met}$ deposits methionine at the intercalary positions only. In eukaryotes, formylation of initiating methionine is not brought about due to the absence of tRNA?er in plants and due to the absence of enzyme transformylase in animals. Initiation in higher organisms will, therefore, take place without formylation. Differences between prokaryotes and eukaryotes in the mechanisms of initiation of translation are summarized in Table.

Initiation in Prokaryotes : Ribosome-mRNA interaction. The initiating codon always functions in combination with a Shine-Delgarno sequence (SD region of mRNA), which has

Differences between prokaryotes and eukaryotes in the mechanism of translation.

Prokaryotes	Eukaryotes
1. Initiating amino acid (methionine) needs to be formylated (therefore, two tRNAs for methionine i.e. $tRNA_f^{met}$ and $tRNA_m^{met}$ are found)	1. Initiating amino acid (methionine) is not formylated (only one tRNA for methionine i.e. $tRNA_i^{met}$ is found)
2. Ribosomes enter the mRNA at AUG codon or at nearby Shine-Delgarno site.	2. Ribosomes enter at the capped 5' end of mRNA and then advance to AUG codon by linear scanning.
3. No initiation factors or co-factors are required for initial contact between ribosomes and mRNA.	3. ATP and a number of protein factors are needed for ribosomes to engage the mRNA.
4. Small ribosomal sub-unit (30S) can engage mRNA before binding of initiator met-$tRNA_f^{met}$.	4. Small ribosomal subunit (40S) binds stably to mRNA only after initiator met-tRNA is bound.

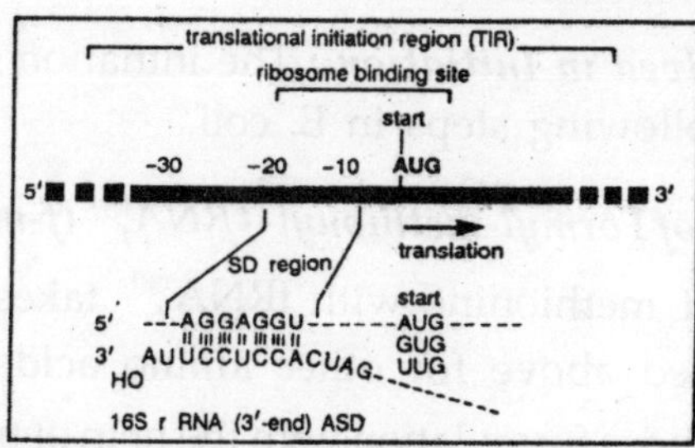

The prokaryotic translational initiation region (TIR), and ribosome binding site (RBS); the Shine-Delgarno region relative to start codon (4-12 bases away) is also shown, which has homology to 3' end of 16S rRNA (ASD region).

complementarity with the 3' end of 16S rRNA (the ASD region) found in 30S subunit of ribosome. SD region is a part of ribosome binding site (RBS), which itself is a part of a longer region called translation initiation region or TIR found in mRNA (Fig.). The TIR helps in binding of mRNA with ribosome, and also accepts formylated methionyl tRNA at the initiating codon.

Interaction of Ribosome-tRNA : In addition to the ribosome-mRNA interaction involving only 16S rRNA, tRNA-rRNA interactions are also known and involved in the tRNA binding to the ribosome. Bases have been identified in the 16S rRNA molecule (G926, m^2G966, G1338, G1401), which are essential for binding of the anticodon stem loop end of tRNA to the P site of the small subunit of the ribosome. Similar bases in the A site will become known in future. The tRNA also interacts with the 23S rRNA of the larger subunit of the ribosome. For instance, the conserved CCA terminus of tRNA interacts with bases in the 23S rRNA at the P site in the larger subunit of the ribosome. The nucleotides of the 23S rRNA involved in this interaction include conserved nucleotides G2252 and G2253 (guanosines at positions 2252 and 2253) in the hairpin loop of this rRNA. It has been shown that C74 (cytosine at position 74) at the acceptor end of a tRNA interacts and forms G-C pair with G2252 in the hairpin loop of 23S rRNA at the P site of the larger subunit.

Steps Involved in Initiation : The initiation of translation involves the following steps in E. coli.

Formation of Formyl-methionyl- $tRNA_f^{met}$ (f-met- $tRNA_f^{met}$): The binding of methionine with $tRNA_f^{met}$ takes place in the manner outlined above for other amino acids. Later, met-$tRNA_f^{met}$ undergoes formylation with the help of transformylase enzyme in the presence of a formate source, formyl-tetrahydrofolic acid.

$$\text{met-tRNA}_f^{met} + \text{formate} \xrightarrow[\text{(formyl-tetrahydrofolic acid)}]{\text{transformylase}} \text{f-met-tRNA}_f^{met}$$

Generation of 30S Subunit and Binding of IF3 : Ribosomes engaged in protein synthesis are released from mRNA as 70S (or 80S) particles. These particles are dissociated into two subunits (30S, 50S) with the help of the protein called initiation factor 3 (IF3). IF3 then binds to 30S subunit to stabilize it, to be ready to bind to mRNA (Figures).

$$70S \xrightarrow{IF3} 50S + 30S;\ 30S + IF3 \rightarrow IF3\text{–}30S$$

Forming of 70S Initiation Complex : Several components, which participate in the formation of 70S initiation complex (70S ribosome + mRNA + f-met- $tRNA_f^{met}$) include the following: (i) 30S and 50S ribosomal subunits, (ii) the initiator tRNA (f-met- $tRNA_f^{met}$), (iii) the TIR of the mRNA, (iv) the three initiation factors (IF1, IF2, IF3) and (v) a GTP molecule. The probable events during initiation are depicted in Figure.

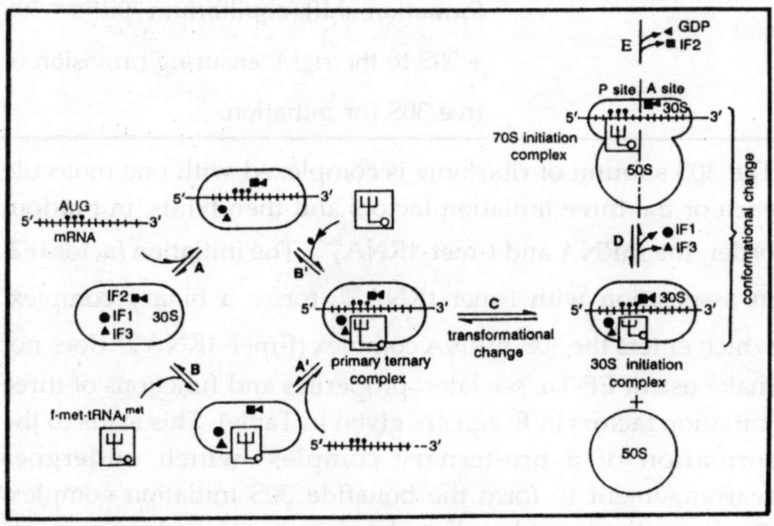

Use of initiation factors IF1, IF2 and IF3 in the formation of initiation complex with 30S subunit and the release of these factors on union with bigger 50S subunit in bacterial systems; A, B, A' and B' are in rapid equilibrium; step C is the most important rate controlling step.

Main properties and functions of three initiation factors in E. coli.

Factor	*Mass (daltons)*	*Function*
IF1	8,100	Stimulates activities of IF2 and IF3; increases the affinity of the 30S subunit for other factors.
1F2	97,300	Kinetic effector of 30S initiation complex formation; favours binding of aminoacyl-tRNAs with blocked 01NH2 groups (e.g. f-met- $tRNA_f^{met}$; positions f-met- $tRNA_f^{met}$ in P site.
IF3	20,700	Kinetic effector of 30S initiation complex formation; shifts equilibrium 70S 5 $\rightleftharpoons$ 30S + 50S to the right, ensuring provision of free 30S for initiation.

The 30S subunit of ribosome is complexed with one molecule each of the three initiation factors and then binds, in random order, the mRNA and f-met- $tRNA_f^{met}$. The initiation factor IF2, in association with f-met- $tRNA_f^{met}$ forms a binary complex, which enters the 30S-mRNA complex (f-met- $tRNA_f^{met}$ does not make use of EF-Tu, see later; properties and functions of three initiation factors in E. coli are given in Table). This leads to the formation of a pre-ternary complex, which undergoes rearrangement to form the bonafide 30S initiation complex. This complex can either dissociate into its original components or will associate with 50S subunit to form a 70S initiation complex. The 30S-50S association is rendered irreversible by the initial ejection of IF1 and IF3 and subsequent release of IF2 after hydrolysis of a GTP molecule. After 30S-50S association, f-met- $tRNA_f^{met}$ (in association with IF2) gets located on the ribosomal P site (P = peptidyl site), so that it can form a peptide bond with the incoming aminoacyl tRNA encoded by the second mRNA codon on the ribosomal A site (A = aminoacyl site). The formation of the first peptide bond marks the transition to the elongation stage of translation.

Eukaryotes Initiation : Initiation of polypeptide chain in eukaryotes is similar to that in prokaryotes, except the following minor differences, (i) In eukaryotes there are more initiation factors and at least ten have been identified in red blood cells. They are named by putting a prefix 'e' to signify their eukaryotic origin. These factors are elFl, eIF2, eIF3, eIF4A, eIF4B, eIF4C, eIF4D, eIF4F, eIF5 and eIF6 (Table). The eIF2, eIF3, eIF4A and eIF4F contain multiple polypeptide chains, but others are single polypeptides; eIF2 and eIF3 are analogous to IF2 and IF3 of prokaryotes. (ii) In eukaryotes, formylation of methionine does not take place, (iii) In eukaryotes, smaller subunit (40S) associates with initiator tRNA known as $tRNA_i^{met}$ (because methionine is not formylated), without the help of mRNA, while in prokaryotes, generally the 30S-mRNA complex is first formed, which then associates with f-met- $tRNA_f^{met}$.

Structure and function of nine initiation factors isolated from reticulocytes in eukaryotes.

Factor	*Structure*	*Function*
	(mass in daltons)	
eIF3	> 500,000 (multimer)	Maintains free 40S subunits; helps binding mRNA with 40S
eIF4F	200,000 (multimer)	Binding mRNA 5' end with 40S subunit; unwinding
elFl	15,000 (monomer)	Assists mRNA binding

Contd..

Factor	*Structure*	*Function*
eIF4A	15,000 (multimer)	Assists mRNA unwinding (also bind ATP)
eIF4B	15,000 (monomer)	Assists mRNA binding & unwinding
eIF6	23,000 (monomer)	Prevents 40S-60S joining
eIF5	150,000 (monomer)	Releasing eIF2 & eIF3;
elF5A	hypusine (unique aa)	joining 60S subunit
eIF4C	15,000 (monomer)	Binding 60S subunit
eIF2	Trimer ($\alpha\beta\gamma$)	Binding met-tRNA
eIF4D	Monomer	Unknown

Following steps are involved in the initiation of eukaryotic polypeptide synthesis : (i) As shown in Figures, GTP binds to eIF2, which increases its affinity for met- $tRNA_i^{met}$ met- $tRNA_i^{met}$ associates with eIF2-GTP complex forming a ternary complex, i.e. met- $tRNA_i^{met}$-eIF2 -GTP. (ii) The ternary complex associates with 40S subunit, to form 43S initiation complex. (The factor eIF2 in mammals and plants like wheat has three subunits, namely α (35,000), β (38,000), γ (55,000); eIF2 α binds to GTP, eIF2y binds to met-tRNA and eIF2 β may be a recycling factor (Table). (iii) The mRNA at its 5' end binds with 43S initiation complex. This reaction depends on eIF3 and the binding of mRNA is assisted by eIF4F, eIF4A, eIF4B and a high energy bond of ATP (Table; Figures). As can be seen in Figure, the association of mRNA takes place on 5' end and not at the initiation codon AUG as in prokaryotes. (iv) After association of the 5' end of mRNA, initiation complex moves towards 3' end of mRNA in search of initiation codon AUG,

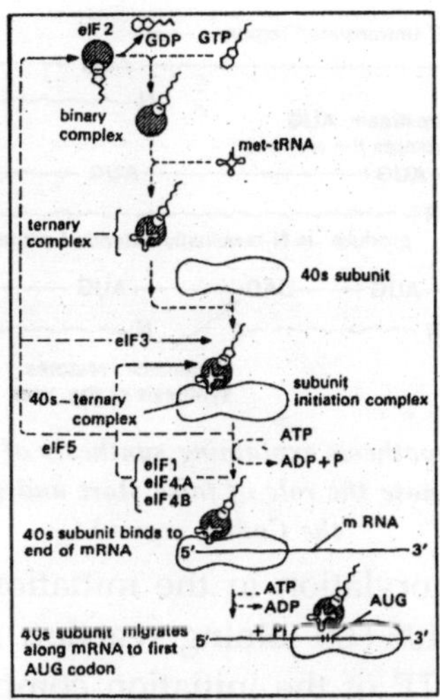

Use of a number of initiation factors in eukafyotes for binding of 40S subunit and mRNA to form a complex. 40S subunit with ternary complex then moves to AUG codon (modified from Lewin's 'Genes').

and then also associates with 60S subunit (you may recall that in prokaryotes, the initiation complex directly attaches to AUG codon and not to the 5' end of mRNA). Association of 60S subunit with the initiation complex requires the factor eIF5, because I helps in releasing eIF2 and eIF3 (60S can not join, if these two factors are not released). eIF2 is released as a binary complex, eIF-2-GDP . The details of the role of eIF2 and its

Structure and function of three subunits of eukaryotic initiation factor eIF2.

Subunit	*Mass (daltons)*	*Function*
α	35,000	Binds GTP
β	38,000	May be recycling factor
γ	55,000	Binds met- $tRNA_f^{met}$

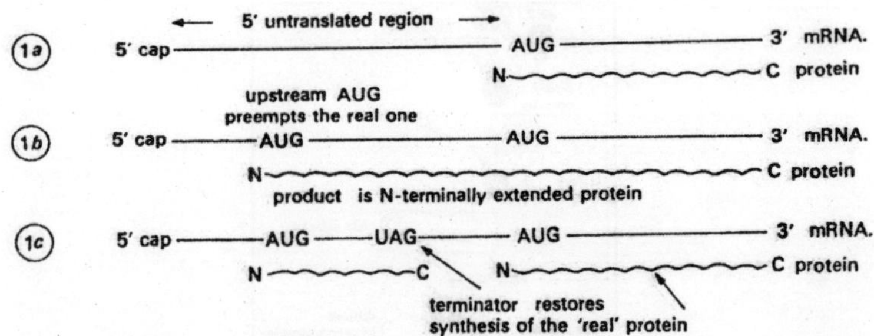

Kozak's scanning hypothesis explaining synthesis of mRNA in eukaryotes (for details see text; note the role of false start and stop signals preceding the Coding region).

reversible phosphorylation in the initiation of translation in eukaryotes. The 40S-60S joining reaction really depends on eIF4C, and the GTP of the initiation complex is hydrolysed, when 60S subunit joins.

It should be realised that the initiation process described above is characterized in prokaryotic system by E. coli and in eukaryotic system by red blood cells, where major work has been done. Variations may by found in other systems.

Translation in the Nucleus in Eukaryotes : In a report published in August 2001, it was shown that 10-15% of protein synthesis in a eukaryotic cell perhaps occurs within the nucleus, although earlier it was believed that translation occurs only in the cytoplasm. If translation in the nucleus is a reality, future research will confirm it and reveal details of this translation system, which does not seem to differ from that in the cytoplasm.

Kozak's Ribosome Scanning Hypothesis for Translation in Eukaryotes : In 1983, Marilyn Kozak proposed a hypothesis for initiation of translation by eukaryotic ribosome. According to this hypothesis, 40S smaller subunit of a eukaryotic ribosome with its associated met-tRNA moves down the mRNA from 5' end, until it encounters the first AUG. At this point, the 60S subunit joins and the translation begins. The 80S ribosome, after reaching termination, releases protein and dissociates in two subunits.

Kozak's hypothesis received support from cases, where one or more AUG codons were present or inserted upstream of the starting codon of a mRNA. In these cases it was found that these extra or false AUG codons were always read, so that if a false AUG codon was in the reading frame of the coding region, it would give an extended protein, and if followed by a termination codon, a small peptide is released (Fig.). However, during early 1990's, Kozak's hypothesis was challenged, when it was shown that in some mRNAs, the ribosome can bind to the initiation codon in the internal ribosome entry sites (IRES) without any prior scanning from the 5' end of these mRNAs. However, these internal binding sites are often the characteristic features of mRNAs that make use of initiation codons other than AUG, and amino acids other than methionine (as discussed below, sometimes codons other than AUG may also be translated into methionine).

Initiation of Translation with Methionine using Codons Other than AUG : Many exceptions to general outline of the mechanism of translation initiation described earlier are now known. For instance, it was shown that in prokaryotes, rarely the codons GUG and UUG were translated into methionine and used for initiation with the help of mutant $tRNA_f^{met}$. Similarly, in eukaryotes, codons ACG, CUG, AUU and AGG are rarely used as initiating codons, and translated into methionine with the help of mutant $tRNA_i^{met}$.

Initiation of Translation without Methionine and/or its tRNA : In the year 1990 and later, cases were also reported, where amino acids other than methionine were used for initiation. For instance in E. coli, UAG codon (a termination codon) was used for initiation with formylglutamine, and without the help of initiator $tRNA_f^{met}$. Later (in 1998, 2000), similar results were reported for the translation of viral RNA in mammalian and insect cells, where the initiating codons were CAA translated into glutamine and GCU/GCA translated into alanine. In most such cases with abnormal translation initiation, the messenger RNA was characterized by the absence of 5' cap and by the presence of an internal ribosome entry site (IRES) associated with a pseudoknot, in the non-coding region, immediately upstream of the codon used for initiation. The pseudoknot may be important for IRES-dependent translation, since any disruption of this pseudoknot has been shown to abolish translation.

Polypeptide's Elongation

In *E. coli,* after the formation of 70S-mRNA-f-met- $tRNA_f^{met}$ complex, elongation of polypeptide chain is brought about by the regular addition of amino acids. At 37°C about 15 amino acids per second are added to a growing polypeptide chain, so that in 20 seconds a protein with 300 amino acids can be synthesized. The following three steps are important in the elongation process.

Binding of AA-tRNA at Site 'A' of Ribosome *(classical vs hybrid state models for translation)* : In earlier classical model, each ribosome had two cavities, in which tRNA could be inserted. These were ' 'P' site (peptidyl site) and 'A' site (amino acyl site). However, later a third cavity called 'E' site (exit site) was suggested (see later for details). It is not known whether f-met- $tRNA_f^{met}$ comes on P site or on A site, but ultimately it has to be present on P site, to make 'A' site available for the next amino acyl tRNA (AA-tRNA).

Another site called 'R' site located on smaller subunit of ribosome was proposed. The presence of 'R' site was confirmed and its role was discussed. It was proposed that 'R' site plays a role in the improvement of accuracy of translation. The aminoacyl tRNA first binds to R site involving codon-anticodon pairing (Fig.). Later aminoacyl tRNA is flipped to A site using energy from GTP molecule (Fig.). During this flipping, tRNA is held only by codon-anticodon pairing. If this pairing is weak due to mismatching of wrong tRNA, aminoacyl tRNA will fly off the ribosome. R site thus provides a correcting measure.

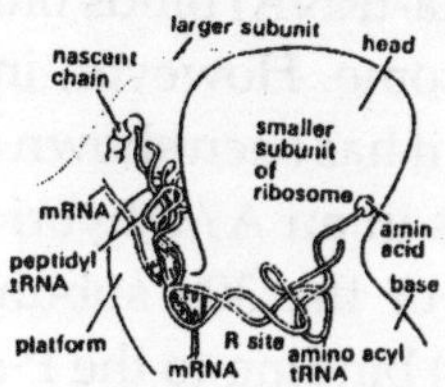

Binding of amino acyl tRNA to R-site, when its anticodon is bound to the next unread codon

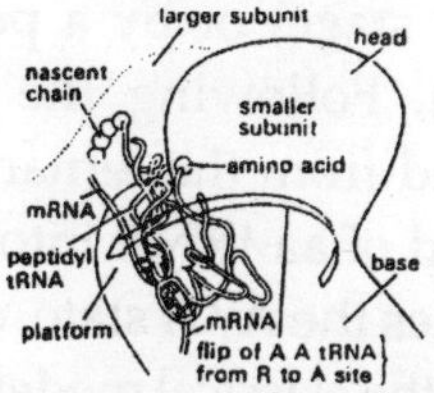

Amino acyl tRNA flips into a position to help its binding on 'A' site

After formation of 70S initiation complex, the next amino acyl tRNA (aa-tRNA) enters 'A' site. Factors responsible for this entry include the elongation factors EF-Tu (u means unstable when heated) and EF-Ts (s means stable when heated). Although aa-tRNA can bind to ribosome even without the help of elongation factors (EF-Tu, EF-Ts), proper binding occurs only in the presence of EF-Tu, which is responsible for placing aa-tRNA on the A site.

The elongation factor EF-Tu (aG-protein) first combines with GTP and changes to an active binary complex (EF-Tu.GTP) which binds with aa-tRNA (but it cannot bind to f-met- $tRNA_f^{met}$, which instead binds with IF_2 to form a binary complex) to form a ternary complex.

EF–Tu–Ts+GTP ⟶ EF–Tu.GTP+EF–Ts
binary complex

EF–Tu.GTP+aa–tRNA ⟶ EF–Tu..GTP.aa–tRNA
(ternary complex)

According to the classical model of translation, the ternary complex (EF-Tu.GTP. aa-tRNA) binds only to the A site in both the subunits of a ribosome. However, in more recent models (hybrid states models), it has been shown that the above ternary complex actually binds in an A/P hybrid state, the anticodon binding to the A-site of the 30S subunit and the CCA end (sequestered by ET-Ttf) binding to the P-site of the 50S subunit as well as to the 30S subunit. At the time of the entry of the ternary complex, the 'P' site is already occupied by f-met. $tRNA^{met}$ (at the initial stage) or by a peptidyl tRNA (at later stages of elongation). Following the GTP hydrolysis, EF-Tu.GDP+P are released from the ternary complex, permitting movement of CCA end of aa-tRNA into A site of the large 50S subunit. This establishes the A/A state, equivalent to the A site binding of aa-tRNA in the classical model of translation. Further, EF-Ts now displaces GDP in the EF-Tu.GDP binary complex and associates with EF-Tu, so that GTP can again associate with EF-Tu to start another cycle for the binding of aa-tRNA. EF-Ts is released at this stage to be used again with EF-Tu.GDP complex (Fig.). EF-Ts is believed to be a molecule which is also described as guanine exchange factor (GEF). Such a GEF is also known in the eukaryotic systems.

EF-Tu.GDP + EF-Ts → EF-Tu.EF-Ts + GDP

EF-Tu.EF-Ts + GTP → EF-Tu.GTP + EF-Ts

In eukaryotes as typified by red blood cells, there are two elongation factors eEFl and eEF2, the former being considered analogous to EF-T (both EF-Tu and EF-Ts) of E. coli and the latter analogous to EF-G (see later in the section). However, eEFl may have several polypeptides in addition to the two which may be analogous to EF-Tu and EF-Ts. The elongation factors in chloroplasts and mitochondria may be completely independent and have not been characterized so far.

Peptide Bond's Formation : This is a catalytic reaction during which a peptide bond is formed between the free carboxyl group (-COOH) of the peptidyl tRNA at the 'P' site and the free amino (-NH_2) group present with amino acyl tRNA, which is available at the A site.

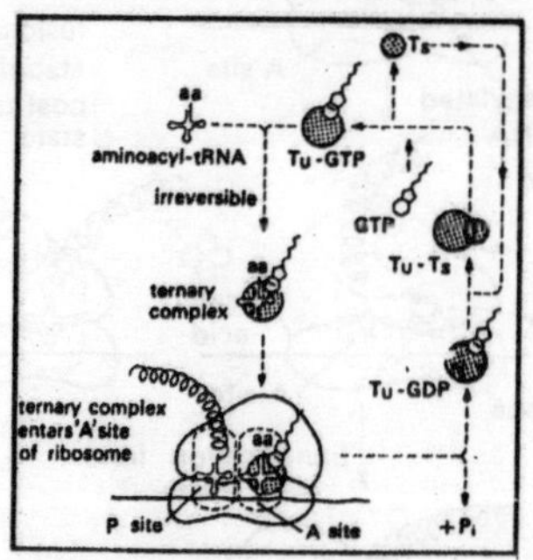

Formation of EF-Tu-GTP-tRNA ternary complex, which helps entry into 'A' site, and the release of EF-Tu-GDP, which forms Tu–Ts.

The 50S rRNA seems to have peptidyl transferase activity, so that the ribosome is described as a ribozyme. The peptidyl transferase reaction involves peptidyl-transferase-coupled translocation of aa-tRNA. It has been shown that the tRNA moves relative to ribosomal proteins S21 and Ll1. A displacement model has been proposed for the peptidyl transferase reaction, which is consistent with the hybrid states model discussed above. According to this displacement model, peptidyl chain remains in a constant position relative to ribosome, while tRNA moves during the peptidyl transferase reaction. After the formation of peptide bond, the tRNA at 'P' site is deacylated and the tRNA at 'A' site now carries the polypeptide.

Translocation of Peptidyl tRNA from 'A' to 'P' Site : The peptidyl tRNA present at 'A' site is now translocated to 'P' site. For translocation of peptidyl tRNA from 'A' site to P site, there are two models available : (i) According to two sites (A, P) model, deacylated tRNA is liberated from 'P' site, and with the help of one GTP molecule and an elongation factor EF-G (called EF-G, since it was isolated as GTP requiring factor; also earlier called transfer factor, TFII or translocase), the peptidyl tRNA

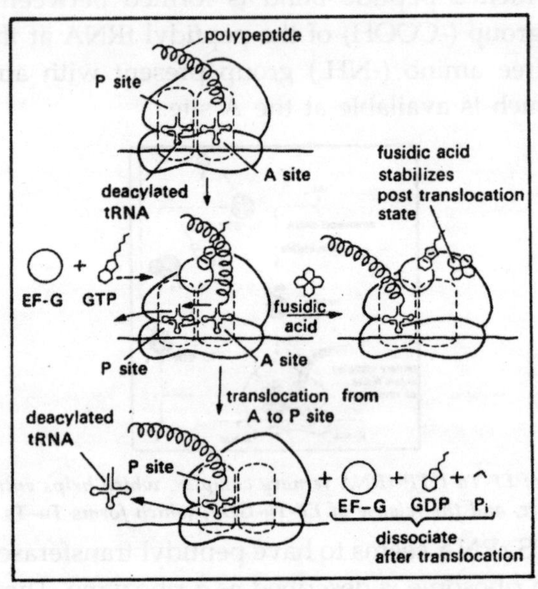

Translocation of peptidyl tRNA from 'A' site to 'P' site with the help of elongation factor EF-G and GTP.

is translocated from 'A' to 'P' site (Fig.). Thus according to this model, tRNA is either entirely in the A site or entirely in the P site, (ii) According to newer three sites (A, P, E) model, initially the aminoacyl end of the t-RNA bound to A-site moves to the P site on 50S subunit at the time of peptide transfer (before translocation), so that initially the newly-formed peptidyl tRNA is in the A/P hybrid state. Only later during translocation, the anticodon end of this tRNA moves from A to P site on 30S subunit so that the peptidyl tRNA is now in the P/P state. Only this later step requires action of elongation factor EF-G. In this model thus, there is an intermediate state, when anticodon end of tRNA is still on A site (on 30S subunit), while the aminoacyl end occupies P site (on 50S subunit; see Fig.). A third tRNA binding site, 'E' (Exit) was also recognized, through which tRNA leaves the ribosome. E site is found mainly on 50S subunit, and deacylated tRNA interacts with it through its CCA sequence at its 3' end.

The movement of deacylated tRNA from P to E site also occurs in two steps : the aminoacyl end moves to E site during peptide bond formation (deacylated tRNA is now in P/E state), and the anticodon end leaves the P site (on 30S) only during translocation (deacylated tRNA is now in E/E state). This three-site model suggests that tRNAs interact mainly at their ends. It also assumes that a tRNA remains bound to ribosome at one or the other of its two ends (aminoacyl end, anticodon end), and does not leave ribosome at both its ends simultaneously (either on A or P site). The-two-step transfer of tRNAs from A to P site and from P to E site could result from reciprocating motions of the two subunits of a ribosome. It means that 50S and 30S subunits move alternately and not simultaneously.

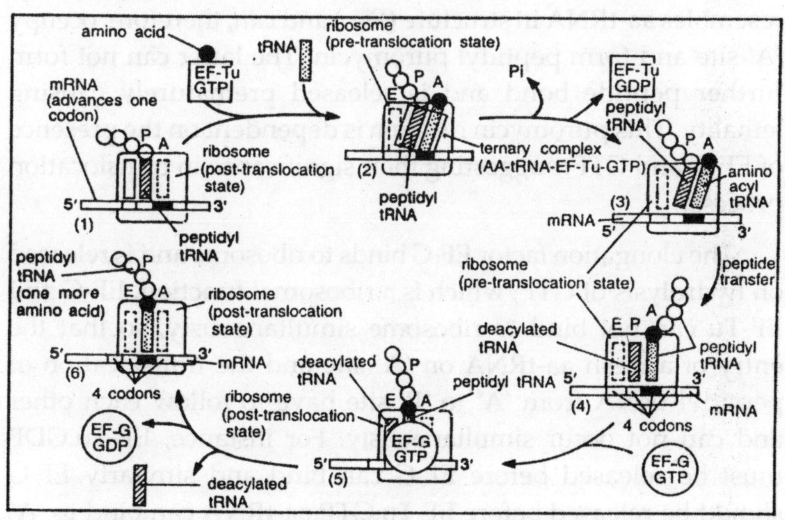

Three site (A, P, E) model for translocation of peptidyl tRNA from A to P site.

The hybrid states model of translation (in comparison of the classical model) suggests the following features of the role of t-RNA in translation (described in greater detail by Green and Noller in a recent review Ann. Rev. Biochem., 1997, 66: 679-716):

(i) the peptidyl moiety remains stationary and only the tRNA moves;

(ii) movement of tRNA (translocation) takes place in two steps -in the first step, both aminoacyl tRNA and peptidyl tRNA move with respect to the larger subunit and in the second step both these tRNAs (along with mRNA) move with respect to the smaller subunit of ribosome.

These features suggest that the transfer RNA binding can have at least six states (A/A, A/P, P/P, P/E, E/E) and perhaps a seventh state, described as A/P* state.

The requirement of EF-G and GTP for translocation was revealed by the use of antibiotic puromycin, which closely resembles aa-tRNA in structure (Fig.) and can, therefore, occupy 'A' site and form peptidyl puromycin. The latter can not form further peptide bond and is released prematurely causing lethality. This puromycin reaction is dependent on the presence of EF-G and GTP, suggesting their significance in translocation process.

The elongation factor EF-G binds to ribosome and is released on hydrolysis of GTP, which is a ribosomal function. EF-G and EF-Tu can not bind to ribosome simultaneously, so that the entry of a fresh aa-tRNA on 'A' site and the translocation of peptidyl tRNA from 'A' to 'P' site have to follow each other and can not occur simultaneously. For instance, EF-Tu.GDP must be released before EF-G can bind and similarly EF-G should be released before EF-Tu.GTP.aa-tRNA can bind to 'A' site. This was confirmed by the observation that if ribosome-

EF-G.GTP complex is stabilized by a steroid antibiotic fusidic acid, so that EF-G is not released after translocation, further

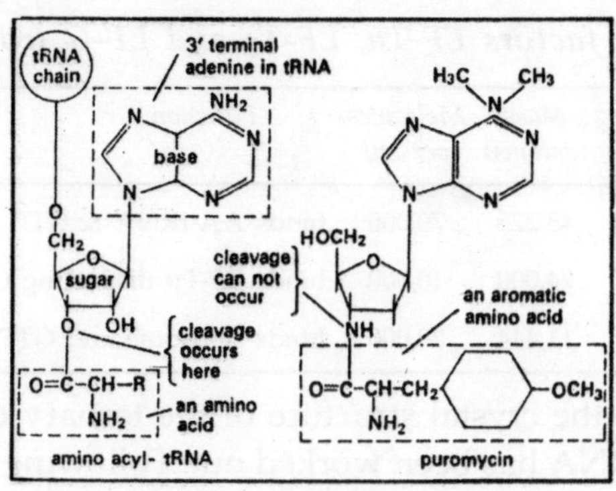

A similarity between amino acyl tRNA and puromycin, so that puromycin can occupy 'A' site, but can not be cleaved due to the presence of N in place of O in RNA.

process of translation stops, due to inability of EF-Tu.GTP.aa-tRNA complex to enter the 'A' site. This exclusion of one elongation factor by the other allows orderly translation in protein synthesis. The details about EF-Tu, EF-Ts and EF-G are given in Table.

In eukaryotes, the elongation factor needed for translocation is called eEF-2. Unlike prokaryotes, a stable complex of eEF-2 with GTP can be isolated and this complex can bind to ribosome.

The above account of elongation of polypeptide suggests that for the formation of one peptide bond, one ATP molecule (for activation of amino acid) and two GTP molecules (one for transfer of aa-tRNA to 'A' site and the other for translocation of peptidyl tRNA from 'A' to 'P' site) are required, However, during 1990's evidence became available, which suggested that the number of GTP molecules for the formation of one peptide bond can be three or even more. For instance, it is suggested that for the transfer of one aa-tRNA, two molecules of the binary complex, EF-Tu.GTP, may be required, so that the total GTP molecules' for one peptide bond will be three (two GTP for transfer of aa-tRNA and one GTP for translocation of peptidyl tRNA).

Elongation factors EF-Tu, EF-Ts and EF-G from E. coli.

Factor	*Genes coding*	*Mass (daltons)*	*Molecules per cell*	*Function*	*Inhibitor*
EF-Tu	*tufA tufB*	43,225	70,000	binds AA-tRNA & GTP	kirromycin
EF-Ts	*tsr*	74,000	10,000	binds EF-Tu displacing GDP	
EF-G	*fus*	11,444	20,000	binds ribosome and GTP	fusidic acid

Recently, the crystal structure of the ternary complex EF-Tu.GTP.aa-tRNA has been worked out. Following are some of the interesting features revealed by this study;

(i) The overall shape of the ternary complex (not the EF-Tu.GDP) resembles EF-G.GDP;

(ii) Three domains of the carboxy -terminal end of EF-G, which have no homology in EF-Tu, together resemble t-RNA and are positioned the same way as tRNA in the ternary complex.

The structural similarity between EF-G and EF-Tu also reflects functional similarity. Both bind to ribosome, when complexed with GTP and have GTPase activity on binding to ribosome. Further, after GTP hydrolysis, both adopt conformations, that have low affinity for ribosomes.

Particularly, similarity has been demonstrated between domain IV of EF-G and the anticodon domain of the ternary complex. During 1997, it was also shown that EF-G dependent hydrolysis of GTP takes place immediately after binding of EF-G.GTP and before the translocation of tRNAs (peptidyl tRNA and deacylated tRNA). These recent studies led to the following conclusions, (i) that GTP hydrolysis by EF-G drives translocation; (ii) that domain IV of EF-G has an important role in coupling GTP hydrolysis to translocation and (iii) that domain IV is essential for the release of EF-G. GDP from the ribosome after translocation.

Polypeptide's Termination

Expiration in mRNA with Stop Codon : Termination of the polypeptide chain is brought about by the presence of any one of the three termination codons, namely UAA, UAG and UGA. These termination codons are recognized by one of the two release factors RF1 and RF2. Of these release factors, RF1 recognizes UAA and UAG and RF2 recognizes UAA and UGA. They help the ribosome to recognize these triplets. The release factors seem to act on 'A' site, since suppressor tRNA capable of recognizing termination codons can compete with release factors by entry at 'A' site. A third release factor RF3 seems to stimulate the action of RF1 and RF2 in a GTP-dependent and codon independent manner (one GTP molecule is hydrolysed during release of a polypeptide, when RF3 stimulates RF1 and RF2. For release reaction, the polypeptidyl tRNA must be present on 'P' site and the release factors help in splitting of the carboxyl group between the polypeptide and the last tRNA carrying this chain. Polypeptide is thus released and the ribosome dissociates into two subunits with the help of ribosome release factor or RRF (IF3 ?). The identity of RRF in eukaryotes is unknown, where ,so far only one release factor is known i.e. eRFl. GTP seems to be necessary for binding of this factor to ribosome. GTP is cleaved after the termination step has occurred which may be a prerequisite for the release of eRFI from the ribosome.

Expiration in mRNA with a Stop Codon (10Sa RNA works both as tRNA and mRNA) : The above termination process cannot be used in the absence of an in-frame termination codon, at the 3'end of mRNA molecule or its segment. This will be the situation in many truncated mRNAs. Under this condition, the ribosome may stall at the end of the message and a newly-discovered 363 nucleotides long RNA called 10Sa RNA (encoded by the E. coil gene ssrA) is bound to this stalled ribosome, initiating a process of degradation of the incomplete polypeptide chain. This 10Sa RNA functions both as a tRNA and a mRNA.

It has properties of tRNA including the G3:U70 base pair (5'-3' base pairing) that is crucial for tRNA-ala identity. Accordingly, it is charged with alanine at its 3'tRNA acceptor-like end. On binding of the charged 10SaRNA to ribosome,

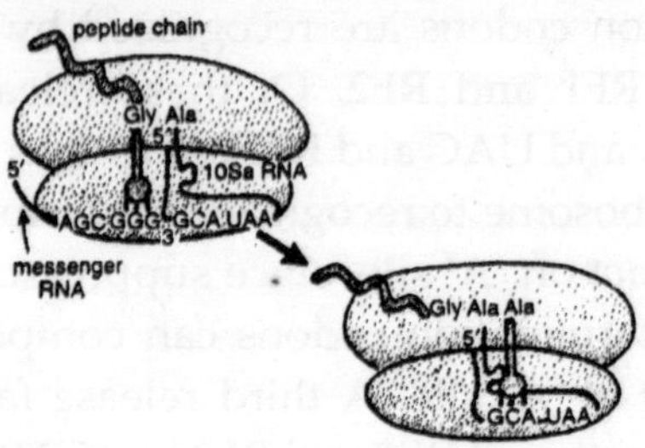

Use of 10 SaRNA for termination in mRNA lacking a stop codon.

alanine gets incorporated to the growing polypeptide chain. The ribosome then uses 10SaRNA as mRNA and decodes a short sequence, adding 10 amino acids (ANDENYALAA) onto the C-terminus end of truncated polypeptide chain and terminates at a UAA codon available in 10Sa RNA (Fig.). The polypeptide is thus released. The C-terminal oligopeptide in this chimaeric chain is recognized by COOH-terminal specific proteases, so that the polypeptide is then degraded. This phenomenon for degradation of incomplete polypeptide remained unnoticed for a long time and was discovered only recently (1994-96). This protein degradation differs from the mechanism, where ubiquitin is added after the translation process is completed and leads to degradation in the proteasome.

Chloroplast's and Mitochondria's Translation : It has been shown that in both chloroplasts and mitochondria, following components of translation apparatus are found : (i) ribosomes specific to organelle, (ii) tRNAs specific to organelle, (iii) other factors for translation (little is known about these factors). Some of these components may be synthesized outside these organelles, and then transported into organelles.

It has been shown that the translation apparatus in chloroplasts and mitochondria differs from that in cytoplasm in eukaryotes in the following respects, (i) Ribosomes in these organelles are smaller in size (70S) than those in cytoplasm (80S), (ii) The tRNAs are specific and differ, the number of tRNAs in mitochondria being 22 as against 55 in cytoplasm, (iii) Initiation of translation takes place by formyl-methionyl tRNA both in chloroplasts and mitochondria, although no formylation takes place in cytoplasm, (iv) Translation in chloroplasts and mitochondria can be inhibited by chloramphenicol, as in bacteria since the 70S ribosomes are sensitive to chloramphenicol and not to cycloheximide; on the other hand, the translation in cytoplasm is inhibited by cycloheximide, since 80S ribosomes are sensitive to cycloheximide. There are other antibiotics like spectinomycin, lincomycin and erythromycin which also inhibit translation in bacteria and in all organelles of eukaryotes. These antibiotics help to stop the translation preferentially either (i) in cytoplasm (by cycloheximide) thus permitting protein synthesis only in chloroplasts or mitochondria or (ii) in organelles (by chloramphenicol) thus permitting protein synthesis only in the cytoplasm.

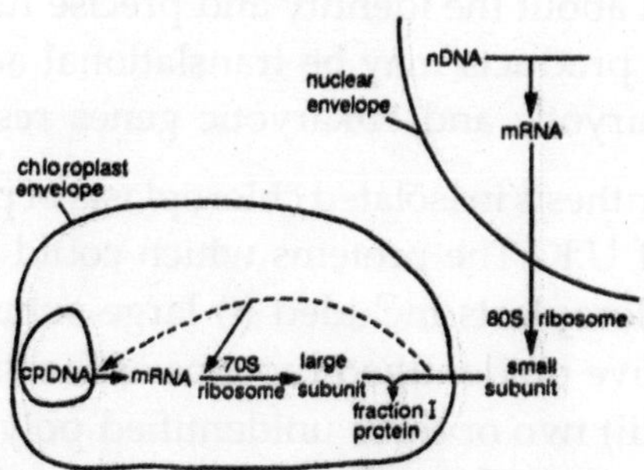

A model of cooperation between nuclear DNA and chloroplast DNA in pea as studied by R.J. Ellis and his colleagues.

Studies on the synthesis of 'chlorophyll-binding proteins', which play a major role in the primary reactions of photosynthesis have also provided new information about chloroplast translation and its regulation. In many plant species, in the absence of light the mRNA of chlorophyll-binding proteins, including those encoded by chloroplast genome, remain associated with thylakoid bound polysomes, but are not translated. Transfer of seedlings to light induces synthesis of proteins, although no increase in transcription has been observed. This suggests that post-transcriptional processes play a key role in light-induced chloroplast gene expression.

It has also been shown that the polycistronic mRNA encoded by the psbB operon in maize, need not be cleaved into tri-, di- or monocistronic forms for successful translation (*psbB* operon = *psbB* + *psbH* + *petB* + *petD*). This suggests that plastid ribosomes can bind directly to internal initiation regions for initiation of translation as in prokaryotes. However, RNA processing does take place in chloroplasts and mitochondria, and the transcripts may differ in their translatability.

There are also nuclear genes, whose products are essential for translation of specific genes in chloroplasts and mitochondria. Several of these nuclear encoded factors interact with the 5' untranslated region of chloroplast massages, but little is known about the identity and precise function of these factors. These products may be translational activators of the type for prokaryotic and eukaryotic genes respectively.

Protein synthesis in isolated chloroplasts of pea was studied by R.J. Ellis of U.K. The proteins which could be synthesized in isolated chloroplasts included (i) large subunit of Fraction I protein, (ii) five unidentified proteins of the internal lamellar system, and (iii) two or three unidentified polypeptides of the envelope. These are only few of a large number of proteins found in chloroplasts. It has also been demonstrated that while the large subunit of Fraction I protein is synthesized under the influence of chloroplast DNA, the small subunit is synthesized in the cytoplasm under the influence of nuclear DNA. Small subunit is then transported to the chloroplast where it is associated with large subunit to give rise to Fraction I protein (Fig.).

Proteins synthesized in mitochondria have also been identified by inhibiting protein synthesis in cytoplasm by cycloheximide. Although, some proteins are synthesized in mitochondria, not all proteins present in mitochondria are synthesized there. There are some mitochondrial proteins which are synthesized in cytoplasm and then transported. There are still others having dual origin, some polypeptides having cytoplasmic origin and others having mitochondrial origin. Some proteins, which are known to be synthesized in mitochondria include the following : (i) Cytochrome oxidase contains seven different kinds of polypeptides, of which three are synthesized on mitochondrial ribosomes, the remaining four being synthesized on cytoplasmic ribosomes. (ii) Adenosine triphosphatase (ATPase) is an important component of mitochondrial membrane, which plays an important role in coupling respiration to ATP formation. There are several polypeptides present in this complex and only two are definitely known to be synthesized on mitochondrial ribosomes. (iii) In mitochondrial ribosomes, although ribosomal proteins are known to be synthesized outside mitochondria and under the influence of nuclear DNA, but rRNA (ribosomal RNA) is transcribed from mtDNA, since both 21S rRNA and 15S rRNA (in yeast mitochondria, 21S rRNA and 15S rRNA occur instead 23S and 16S rRNAs) hybridize with specific regions of mtDNA.

Alternation and Enfolding of Freed Polypeptide

Alternation of Freed Polypeptide : After translation, the released polypeptide is modified in various ways. An enzyme deformylase removes the formyl group of first amino acid methionine. (Fig.). Due to the action of certain other enzymes, exo-aminopeptidases, amino acids may be removed from either the N-terminal end or the C terminal end or both. Due to this modification, therefore, the N-terminal amino acid may not be methionine in some cases (Fig.), although in protein synthesis, the first amino acid is always methionine. The polypeptide chain singly or in association with other chains also folds into a tertiary structure. This problem of protein folding is sometimes described as 'Second Half of the Genetic code'.

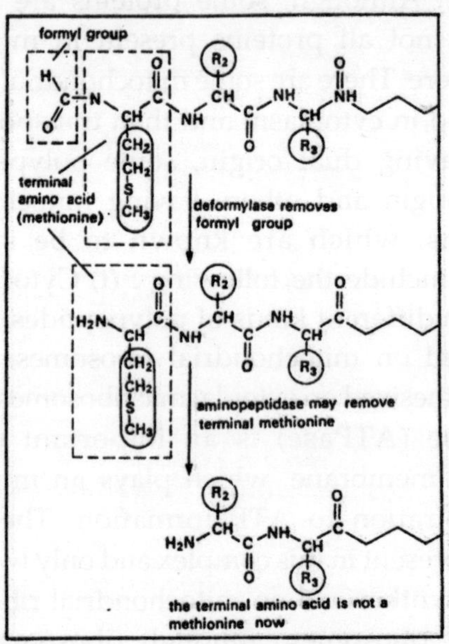

Removal of formyl group from released polypeptide and a subsequent removal of methionine, so that the second amino acid becomes terminal (modified from Lewin's 'Genes').

Inteins and their Excision from Polypeptides : Split genes, which consist of intron and exon sequences. We have also discussed the mechanisms used for the excision of intron sequences from pre-mRNA to give mRNA. A parallel situation sometimes exists in polypeptides that are released after translation. In other words, some polypeptides may contain sequences called inteins, that are destined to be spliced out post-translationally by a proteolytic cleavage and ligation process. The sequences that flank inteins and are retained are called exeins. Inteins appear to autocatalyse their own excision.

Some of them are site-specific endonucleases, which can insert each a copy of its own DNA into its integration site in an inteinless allele. Thus, at the DNA level, like introns, intein-encoding DNA sequences have also been shown to be 'mobile genetic elements'. Till about the middle of 1998, more than fifty inteins were already known from genomes of bacteriophages, mycobacteria, cyanobacteria, archaebacteria, yeast and chloroplast of a red alga. Some of the proteins, known to have inteins, include DnaB, RecA, DNA polymerases. Intein encoding DNA sequences have also been shown to coexist with introns in the same genes. In an evolutionary sense, inteins are believed to be very ancient and might have been lost in other more highly evolved polypeptides. Alternatively these may represent molecular parasites, that invaded the corresponding genes in the past and merely needed to ensure the transmission of their genes. The shortest known intein is 150 amino acids long and is found in the DmiB protein of chloroplasts of the red alga Porphyra purpurea: It lacks the endonuclease-like motifs found in all other inteins, which range in their length from 360 to 538 amino acids.

Chaperones in Enfolding and Gathering of Proteins : Molecular chaperones represent a diverse class of protein molecules, which play a significant role in folding of individual polypeptides and in assembly of multimeric proteins, but are' not themselves components of the final functional structures. These chaperones function by inhibiting incorrect molecular interactions, which are often possible in their absence (word chaperone, rather chaperon is used in English to describe 'a person' usually a married or elderly woman who for the sake of propriety, accompanies a young unmarried lady in public, as guide and protector). Some molecular chaperones (but not all of them) are also stress proteins, so that it is possible that all stress proteins act as molecular chaperones. Another term 'chaperonin' is sometimes used interchangeably with 'molecular chaperones' , but refers to only one of the different classes of

Some families of molecular chaperones.

Class	*Members*
1. Nucleoplasmins	Nucleoplasmin Protein XLNO38 Protein Ch-NO38 Nucleoplasmin S
2. Chaperonins	Chaperonin 60 (cpn60) 1. Bacterial groEL (E. coli) 65kDa common, antigen (inycobacteria) 2. Plastid : rubisco subunit-binding protein 3. Mitochondrial : hsp 60 mitonin HuCha60 4. Eukaryotic cytosol : TCP-1 Chaperonin 10 (cpnlO) 1. Bacterial groES 2. Mitochondrial
3. Heat shock proteins 70 (hsp70)	Hsp68, 72, 73; DnaK; clathrin uncoating ATPase; BiP; grp75, 78, 80; hsc 70, KAR2; SSA1-4; SSB1; SSC1; SSD1
4. Heat shock proteins 90	Hsp83, 87; HtpG
5. Signal recognition particle	SRP
6. Subtilisin prosequence	
7. Alphalytic factor prosequence	
8. Ubiquitinated ribosomal proteins	
9. Trigger factor	
10. Sec B protein	
11. Pap D protein	

molecular chaperones. (Table). Many chaperonins are ATPases, whose function was demonstrated by in vitro folding and assembly of RUBISCO, which required groEL and groES proteins of E. coli. Another chaperonin encoded by a nuclear gene is hsp60 (heat shock protein) which is translated in the cytosol and imported into the mitochondrion, where it is folded and assembled into a tetradecameric structure. Some chaperonins can induce their own folding and self assembly (e.g. groEL protein); there are others (e.g. hsp60), which can fold and assemble only in the presence of other pre-existing functional chaperonins.

Translocation of Proteins in Across Membranes : The proteins synthesized in the cell have to be translocated to appropriate locations within the same cell or other cells. This process is called protein sorting or protein trafficking and involves interaction of cell membranes either with some part of a newly synthesized protein or with a protein sequence added to this new protein. These protein sequences interacting with membranes are called signal sequences or transit peptides, or leader sequences, which are recognized by receptors located within the membrane. The transfer of proteins may be coupled with translation— co-translational transfer (proteins synthesized on ribosomes attached to endoplasmic reticulum = rough ER) or may take place after the translation— post-translational transfer (proteins synthesized on free ribosomes). The signal sequences and the mechanism of intracellular protein trafficking were discovered by Gunter Blobel, who was awarded the 1999 Nobel Prize in Medicine or Physiology.

Protein translocation across membranes has been classified in the following two major systems based on whether these are transported out of the cytosol or imported into the cytosol; (i) export system and (ii) import system. Export system includes transport of proteins out of the cytosol or cytosol-like compartments into extracytosolic compartments including bacterial periplasm (across bacterial plasma membrane), lumen of the endoplasmic reticulum (from the cytosol), thylakoid lumen (across the internal membrane of chloroplasts) and to the mitochondrial inner membrane (from the cytosol). The import system, on the other hand, includes transport of proteins into the cytosol or cytosol-like compartments (eg. stroma of

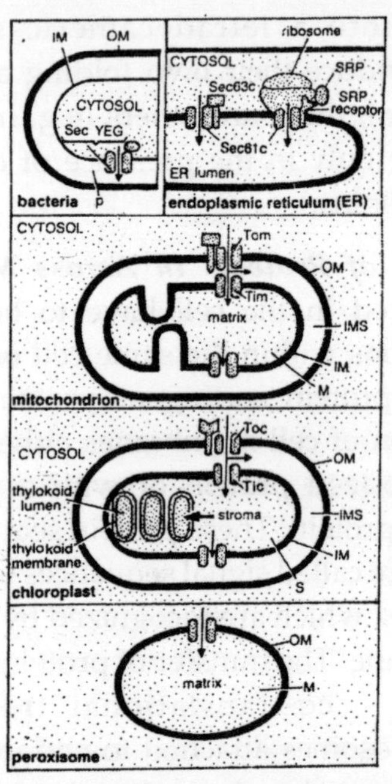

A comparison of protein trafficking among bacteria, eukaryotic cells and eukaryotic organelles.

chloroplasts and matrix of mitochondria and peroxisomes). Thus the mitochondria and chloroplasts have export as well as import systems, while bacterial cells and ER have only export system, and the peroxisomes have only an import system. This does not mean that no macromolecules leave ER, but they generally do so by vesicular traffic. Further, besides these export and import systems, there may be special systems for transport of special proteins. Similarly, nuclear pores allow nucleocytoplasmic transport. Since, these special systems are discussed elsewhere in this book, only the general export and import systems will be discussed in this section. In these systems following are the general features applicable to all export and import transport systems: (i) the protein to be transported carries an -NH_2 terminal signal sequence, due to which it is recognized by some cytosolic factors; (ii) cytosolic factors (e.g. chaperones) recognize membrane receptoı and deliver the protein to the target membrane; (iii) the protein usually remains partly unfolded during transport; (iv) the signal sequence opens a hetero-oligomeric transmembrane channel that is gated both across and within the plane of the membrane (see Fig.); (v) the channel is coupled to a motor protein that is driven by the hydrolysis of ATP or GTP, and transports (by push or pull) the protein across the membrane; (vi) the protein folds on the trans-side of the membrane with the help of chaperones or folding enzymes, (trans-side means the side of the membrane away from the cis surface, where transport is initiated). These general features of export and import systems will be briefly discussed.

Targeting Signals at the NH_2 Terminus of Transported Protein: On the -NH_2 end of each protein destined to be transported, a signal sequence is always found, which is either. cleaved off, once the protein reaches the trans-side of the membrane or may function as signal anchor sequence that permanently anchors the protein within the membrane generating a transmembrane protein. This signal sequence is also called leader sequence, targeting sequence, transit sequence or a pre-sequence. The signal sequence is hydrophobic in export systems and hydrophilic in import systems. Some of the general features of these export and import signal sequences are shown in Figure.

Export Signal Sequences : In E. coli, the export signal sequence acts in two steps : (i) In the first step, it may retard folding of protein facilitating its binding to cytosolic chaperone,

SecB, which then interacts with SecA subunit of transport machinery (in the plasma membrane). The SecB and SecA thus deliver the protein to SecYEG complex in the membrane (SecYEG is a complex of SecY. SecE and SecG). (ii) In the second step, the signal sequence does not act as a targeting signal, but only as a folding inhibitor. It interacts with, and activates the membrane-associated export machinery and facilitates the recognition of protein by the transport machinery.

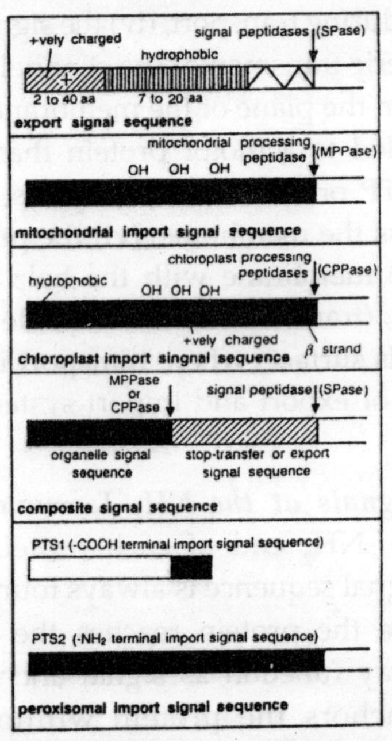

Different signal sequences known for transport of proteins across membranes.

The transported protein may either cross the membrane to reach either the trans side of a lumen, or it may become a transmembrane protein either (i) due to a 'signal anchor sequence' as mentioned above or, (ii) due to sequential function of a 'signal sequence' and a 'stop transfer sequence'. In the second mechanism, the signal sequence initiates translocation and the stop transfer sequence then triggers a lateral opening of the translocating channel, thus arresting the translocation.

stop transfer sequence then becomes the transmembrane B anchor of an integral membrane protein.

Sequences of Import Signal : Import signal sequences (20-35 residues long) help in transport OF the proteins into the mitochondrial matrix or the chloroplast stroma. They are more hydrophilic than export signal sequences and are also rich in hydorxylated amino acids. Composite signal sequences are also known, where an import signal sequence and an export signal sequence (or a stop transfer sequence) are fused and function in a sequential manner one after the other. The import of proteins into thylakoids or to the mitochondrial intermembrane space is often affected by such a composite signal sequence. For instance, in chloroplast, the import signal sequence directs the protein to the stroma, where it is cleaved off and the export signal sequence then transports the protein to the thylakoids. Similarly, in mitochondria, the import signal sequence directs the protein towards mitochondrial matrix and a stop transfer sequence then arrests translocation either in the outer or in the inner membrane, so that the protein then becomes a transmembrane protein. The proteolytic cleavage of a protein stopped in the inner membrane may generate a soluble protein of intermembrane space.

Cytosolic Chaperones and GTPases : For the purpose of transport, a protein should remain at least partly unfolded. Some small proteins (eg coat protein of phage Ml3) remain unfolded and traverse from ribosome to the target membrane without any help from cytosolic factors. However, large proteins with signal sequences require for their transport the assistance of cytosolic chaperones or other similar factors. Later, interaction with membrane receptor, leads to release of cytosolic factors (assisted by ATP or GTP) and directs the protein's signal sequence into the translocation channel (also called translocon). Only, rarely a protein remains folded in cytosol and gets unfolded by the membrane-linked protein transport system.

Chaperones also differ in their specificity. For instance, Hsp70s interact with a wide spectrum of non-native proteins, while SecB and MSF (mitochondrial import-stimulating factor) only interact with proteins destined for transport into or across a membrane. Other factors are also needed. Release from SecB needs SecA and release from Hsp70 and Dnak requires interaction with co-chaperones DnaJ and GrpE accompanied with ATP hydrolysis. A higher level of complexity and regulation is involved in case of cytosolic RNPs termed signal recognition particle (SRP) (see later in this Chapter).

In *E. coli,* proteins can be transported to transport machinery in plasma membrane by any of the two pathways (Fig.), one mediated by SecB and SecA and the other mediated by SRP and FtsY (both are GTPases; FtsY is known as SRa in eukaryotes). In the first pathway cytosolic SecB (a chaperone) binds the protein and delivers it to membrane•bound SecA. In the second pathway, SRP (4.5S RNA + Ffh protein; Ffh a GTPase) binds to the nascent secretory protein and the protein-SRP-ribosome complex then binds to FtsY (GTP dependent reaction). SRP and FtsY increase the efficiency of transporting the protein.

SRP dependent and SRP independent translocations of proteins, destined to endoplasmic reticulum (ER), occur in eukaryotes also, where SRP dependent translocation is coupled with translation. Proteins can also be translocated to ER in a SRP independent manner as shown in yeast. It makes use of Hsp70 and DnaJ homologue Ydjlp, which together help in transporting proteins to mitochondria and other organelles.

The synthesis of most of the mammalian membrane proteins and secretory proteins begins on cytosolic free ribosomes. These proteins are co-translationally targeted to the membrane of the ER. They are recognized and selected with the help of three GTPases (SRP54 belonging to SRP; SRα and SRp belonging to SRP receptor). These GTPases work as adapters between the

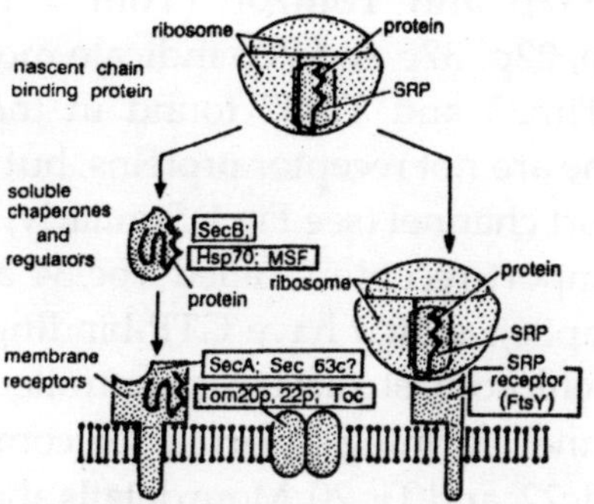

Transport of proteins to the transport machinery in plasma membrane by two alternative pathways.

translation machinery in the cytosol and the translocon (translocation apparatus) in the ER membrane. The translocon is mainly composed of the heterotrimeric Sec61bp. The GTPases bind to GTP and the GTP-binding activity of at least one of them (SRP54=54kDa) is modulated by the ribosome itself. SRP54 (equivalent of Ffh of E. coli)) is a subunit of signal-recognition particle (SRP), which is a cytosolic ribonucleoprotein (RNP), and identifies the ribosomes engaged in the synthesis of nascent chains bearing ER-directed signal sequences. SRP consists of RNA and six polypeptides, SRP54 being the most important. The amino-terminal domain of SRP54 contains GTPase activity and the carboxy - terminal domain binds to signal sequence and SRP - RNA.

Membrane Receptors : In eukaryotes, the membranes involved in translocation carry receptors that recognize the signal sequences bound to cytosolic factors. These receptors are integral membrane proteins with domains that are exposed to the cytosol and their function may be ATP or GTP dependent. SRP receptor, which recognizes SRP is involved in SRP-dependent protein transport to the ER, but no receptor for SRP-independent transport has been identified so far.

Mitochondria from yeast carry in their outer membrane 3 or 4 proteins functioning as receptors. These include Tom20p,

Tom22p, Tom37p and Tom70p (Tom = transport outer membrane; 20p, 22p, 37p and 70p indicate molecular masses). The proteins Tim23 and Timl7 found in the mitochondrial inner membrane are not receptor proteins, but are constituents of protein import channel (see Fig.). Similarly, the chloroplasts have protein-import receptors called Toc 34 and Toe 86 (also called Iap34, Iap86), which have GTP-binding motifs, so that the binding to chloroplast surface is controlled by GTP. Toc75 forms the channel, forming a barrel; Tic complex consists of TicllO, Tic53, Tic22, and Tic 20. More details about the transport of proteins from cytosol to mitochondria and chloroplast have already been discussed.

Transmembrane Protein Channels and Motors : The protein, destined to be transported through a membrane, actually moves through a hydrophilic, hetero-oligomeric transmembrane channel composed of integral membrane proteins. The movement of proteins across the membrane and within the plane of the membrane is facilitated by the following structures: (i) the signal sequence helps in gating (opening) the channel across the membrane, (ii) the stop-transfer sequence or signal-anchor sequence helps in gating the channel within the plane of the membrane; (iii) translocation of the protein is powered by a peripherally-attached protein translocation motor that hydrolyzes ATP or GTP and then pushes or pulls the protein across the membrane. In bacteria and mitochondria, these motors are aided by a transmembrane electrochemical gradient. However, protein transport into thylakoids in chloroplasts can be driven by a pH gradient also.

Several transmembrane protein channels have been characterized. Some of them include the following : (i) In E. coli plasma membrane, the transmembrane channel is a heterotrimer composed of the integral proteins SecY, SecE and SecG, hence the name SecYEG; this acts in concert with SecA and the membrane proteins SecD and SecF. (ii) The channel across the ER of mammals and yeast is a Sec61 complex having three subunits similar in sequence to the three subunits of SecYEG complex found in bacteria, (iii) In the inner membrane of mitochondria, the import channel consists of integral proteins Tim23 and Tim 17 and perhaps two additional subunits. (iv) In chloroplasts, the transport channel across the membrane consists of two integral proteins of the inner envelope, namely lap 100 and Iap36.

Protein transport channels are not active transporters, but function like conduits (tubes) that resemble F_o portion of F_1F_0-ATPase. In this F_1F_0-ATPase, active proton transport is affected by coupling the gated proton channel F_0 to the peripheral proton pump F_1 which is powered by the hydrolysis of ATP. Similarly, as indicated above, a i peripheral attached motor protein is powered by the hydrolysis of GTP or ATP thus

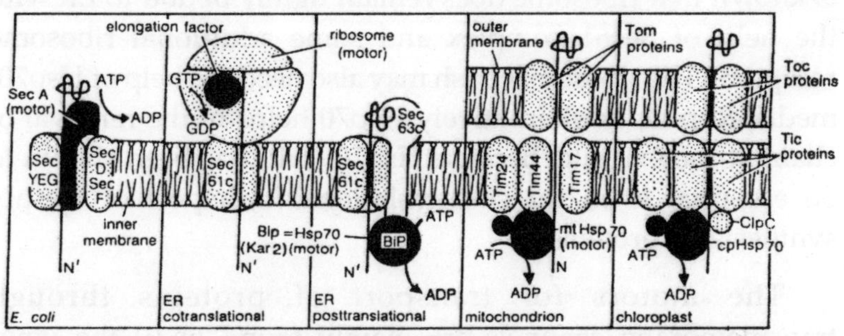

Transmembrane protein channels and motor proteins in E. coli, and endoplasmic reticulum (ER), mitochondria and chloroplast of higher eukaryotes.

pushing or pulling the protein through a protein transport channel-across the membrane. Following are some examples of these motors, (i) Hsp70 (also termed Kar2 or BiP) is found on trans (inner) side of the membrane and pulls the protein into ER as a part of post-translational import; (ii) SecA is bound on the cis (outer) side of the membrane and pushes the protein across the bacterial plasma membrane or chloroplast thylakoid membrane; (iii) ribosome bound to cis side of membrane may also work as a motor and may provide GTP-powered push (Fig.)

In the above discussion, it may be noted that in bacteria, the protein is pushed through the channel, while in ER and mitochondria it is pulled. It may be due to lack of ATP on the trans side of the bacterial plasma membrane, so that one may assume that the bacteria have put the motor where the fuel (ATP) is available.

The ribosome is also believed to work as a motor, so that if the ribosome is fixed to the membrane independent of the translocating polypeptide chain, then the push can be exerted by the GTP-powered polypeptide elongation machinery. Thus, the growing polypeptide is extruded from the ribosome tunnel and traverses through the transmembrane protein channel. It is known that ribosome does remain tightly bound to ER with the help of Sec61 complex and some additional ribosome receptors. The ribosomal push may also need the help of Hsp70-mediated pull, or alternatively Hsp70 helps in the removal of chain at the exit site, because Hsp70 (BiP) has been shown to be essential even for co-translational transport of freshly synthesized proteins.

The motors for transport of proteins through transmembrane channels are mainly found in (i) the inner membranes of chloroplasts and mitochondria, (ii) ER, (iii) thylakoid and (iv) bacterial plasma membrane. They seem to be absent on the outer membranes of mitochondria and chloroplasts. Despite this, proteins are transported through these outer membranes to reach internal compartments. It is suggested that a protein with signal sequences is driven across the membrane by association with one or more binding sites on the inner side of the outer membrane. These binding sites may be provided by Tom22p in mitochondria and by Hsp70 in the chloroplasts. Thus the protein is translocated across the outer membrane of chloroplasts and mitochondria by a relay of binding sites on both faces of the membrane. (A similar cascade of binding sites is involved in the movement of proteins through the nuclear pore complex). Once the protein reaches the inter-membrane space after crossing the outer membrane, it may be captured by the transport system of the inner membrane, where an ATP-linked motor helps further in the translocation process. In mitochondria, this job is done by the electric potential across the membrane. This potential pulls the positively charged signal sequence across the inner membrane.

The components of the machines involved in the export and import systems of translocation differ very much and are listed in Table. The only exception is Hsp70, which is not only involved in both these systems, but also in a variety of other aspects of protein folding.

Components of export and import systems of translocation in E. coli, ER, chloroplasts, mitochondria and peroxisomes.

Components of transport system	Type of system					
	Export systems			Import systems		
	E. coli	E.R.	Chloroplast (thylakoid)	Mitochondrion	Chloroplast	Peroxisome
Cytosolic factors (chaperones or regulators)	Trigger factor; Sec B	NAC	—	—	—	Pas10p (Pas 8p); pas 7P
	Dnak	Hsp70	Hsp70	Hsp70	Hsp70	—
	SRP-GTP, FtsY-GTP	SRP-GTP	SRP-GTP	MSF–ATP	—	—
Membrane receptors	SecA	—	cpSec A	Tom 20p-Tom 22p	Toc86,Toc34	Pex13p,pex14p
	Fts Y	SRα, SRβ	—	Tom 37p-Tom 70p		
Channels	Sec Y	Sec 61p (Sec 61α)	Sec Y	Tom7, Tom6, Tom5	Toc75	
	Sec G	Sec 1p (Sec 61β)	—	Tim 17p, Tim23p		
	Sec E	Sec 1p (Sec 61γ)'	—	—	—	—
		Sec 63p, 62p, 71p, 72p	—	—	—	—
	Sec D, Sec F			Tim 44p	?	—
Translocation motors	Sec A	—	cpSec A	—	—	?
	—	Ribosome	—	—	—	—
	—	Kar2p (BiP)	—	mtHsp70	cpHsp 70	—
Outer membrane channels of organelles				Tom 40p, 6p, 7p, 8p	Toc75; Tic110, 55, 22, 20	

Protein Folding at the End of Transport (Chaperon and Folding Catalysts) : Once the protein confers out of the transport channels, it is folded on the trans side of the membrane. This folding is mediated by a battery of chaperones and other folding helpers, whose function may be specific for a protein or may be general in nature. In the stroma of chloroplasts and matrix of mitochondria, homologues of bacterial GroEL and GroES (termed Hsp60 and Hspl0, or Cpn60 and Cpn10) are found, which help in the folding of many but not all of the imported proteins. They also help in the folding of proteins Synthesized within these organelles. Proteins exported to periplasm in bacteria are refolded with the help of several chaperones and by at least one proline rotamase. ER lumen also contains several chaperones (eg., BiP, Hsp90, calnexin), which facilitate folding of all imported proteins.

Import of Proteins in Peroxisomes : Most proteins imported to peroxisomes carry a COOH-terminal Ser-Lys-Leu tripeptide motif collectively called PTS1 (peroxisome-targeting signal). A few other imported proteins carry a transient NH_2-terminal signal (termed PTS2). PTS1 signal is recognized by a protein Pas8p or its homologue Pas 10p (depending on which yeast species is involved). Similarly, PTS2 is recognized by Pas7p. Pas8p (or Pas 10p) is recognized by peroxisomal integral protein Pas20p. The system thus is comparable to other systems using cytosolic factors, coupled to membrane receptor, but no channels or motors are known yet.

9

Protein's Separation

Living cells are largely made up of proteins (including enzymes), which make more than half of their dry weight. These proteins play an important role in all biological processes that are carried out within a cell. We discussed the details of the genetic code, outlining the manner in which information about the structure of proteins to be synthesized within a cell is stored in another macromolecule, the DNA. We described the apparatus used for protein synthesis and the mechanism of protein synthesis. We described how this synthesis of proteins is regulated within a cell according to an organism's needs. It is thus apparent that the study of the structure and regulated synthesis of proteins is an important component of any molecular biology course, and is no less important than the study of the structure and function of the cell's master molecule, the DNA. However, an aspect of proteins that has not been discussed in any of this book is the techniques used for extraction, purification and identification of proteins. The major steps involved in extraction and purification of a protein from cells are shown schematically in Figure . These techniques will be briefly described in this chapter.

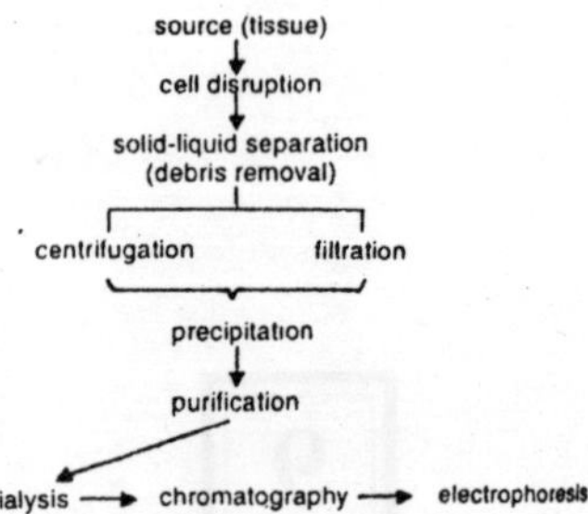

Different steps involved in separation and purification of proteins/enzymes.

Protein's Extraction by Cell Disruption

The proteins are extracted from the cells by breaking the cell walls, thus disrupting the cells causing cell lysis and releasing cell's constituents. There are three main methods for release of intracellular proteins from microorganisms or from cultured eukaryotic cells. These methods include:

(i) enzymatic disruption of cells,

(ii) chemical cell lysis, and

(iii) physical disruption of cells.

Disruption of Cells Enzymatically

Lysozyme is an enzyme, which is produced commercially from hen's egg white, and hydrolyses P-l, 4 glycosidic bonds in the mucopeptide of bacterial cell walls. Gram-positive bacteria and Gram-negative bacteria differ in their susceptibility to lysozyme. In Gram-positive bacteria, lysozyme enzyme alone causes cell lysis, although final rupture of cell wall often depends upon osmotic effects of the suspending buffer. In Gram negative bacteria, however, EDTA needs to be added to lysozyme to achieve cell lysis. This enzymatic method, however, is seldom used for large-scale extraction of a protein, due to high cost of the enzyme lysozyme. It has, however, been used for large-scale release of the enzyme, acryl acylamidase from *Pseudomonas fluorescens*.

Chemical Methods of Cell Lysis

There are at least two chemical methods for cell lysis :

(i) *Alkali Treatment* : This method has been successfully used for large-scale extraction of bacterial proteins. In some cases, exposure of cells to pH between 11.0 and 12.5 can provide the alkali medium for extraction of proteins,

(ii) *Use of Detergents* : Both ionic and non-ionic detergents can be used for protein extraction. The ionic detergents used can be anionic (e.g. sodium lauryl sulphate, sodium cholate) or cationic (cetyltrimethylammonium bromide = CTAB), and non-ionic detergents include Triton X -100 (or X-450) or even Tween. However, ionic detergents are more reactive and can lead to denaturation of many proteins. Presence of detergents, however, affect subsequent purification, which can be overcome by ion-exchange chromatography or ultrafiltration.

Disruption of Cells Physically

Physical methods for cell disruption include the following four methods :

(i) *Osmotic Shock* : This method involves washing the cells in a buffer solution to free them from growth medium, and then suspending them in 20% buffered sucrose. After equilibration, the cells are harvested and rapidly resuspended in water at about 4°C. The method is suitable for release of hydrolytic enzymes and binding proteins from the periplasmic space of Gram-negative bacteria like *E. coli*. The method is particularly suitable for extracting recombinant proteins secreted in the periplasm and has been successfully used for extracting recombinant human growth hormone (rHGH) from *E. coli* cells,

(ii) *Grinding with Abrasives :* Initially this method involved grinding of cell pastes in a mortar with an abrasive powder (e.g. glass, alumina or kieselguhr). Machines have now been developed for the same purpose. One such machine developed by Dynomill (Switzerland) consists of a disintegration chamber (600ml to 250 litres) with glass beads and a number of fixed and rotating impeller discs. The machine can handle from 5 kg to 400 kg of bacteria per hour,

(iii) *Solid Shear :* This method involves the extrusion of frozen cell material through a narrow orifice at high pressure and an outlet temperature of about -20°C. A machine designed for this purpose can handle up to 10 kg bacterial cell paste per hour,

(iv) *Liquid Shear :* The cells in a liquid suspension are passed through a restricted orifice under high pressure (as also in "solid shear"). A homogenizer used in dairy industry can also be used for this purpose (e.g. Manton Gaulin Homogenizer of APVLtd, UK). Liquid shear is the principal choice for large-scale disruption of microbial, yeast or fungal cells for extraction of proteins.

Starting Purification - Removal of Debris : After cell disruption, solids (cell debris) are separated from liquid, the latter containing the proteins/enzymes in solution. This can be achieved either by centrifugation or by membrane filtration.

Centrifugation : Batch centrifuges, varying in capacity from less than 1 ml to several litres, and capable of applying centrifugal force of up to 100,000g are used for removing cell debris, although the cell debris precipitates even at 20,000g. In these batch centrifuges, continuous-flow centrifugation is used with the help of a 'hollow bowl centrifuge' or "disc centrifuge' or a 'multi-chambered bowl centrifuge'. In all these centrifuges, the cell debris or particulate matter is thrown to the side of the bowl and the sedimented material is collected. Sometimes, a cellulose-based Cell Debris Remover (CDR) produced by Whatman Ltd. (UK) is used to help protein/ enzyme recovery.

Membrane Filtration : Tangential or cross-flow filtration is used, where the extract flows at right angle to the direction of filtration. This technique has been put to only a limited use.

Protein's/Enzyme's Precipitation

Enzymes and other proteins are highly charged molecules, and can be precipitated with appropriate charge neutralizing chemicals. Once their charges are broken, they form aggregates and settle down as precipitate. When, an acid or base is added, the protein can be brought to its isoelectric pH. At this pH, there is no net charge on enzyme molecules and electrostatic repulsion between them is low, so that they tend to aggregate. Therefore, adjusting the pH to isoelectric point of a protein, causes its precipitation.

Acids and bases, however, often inactivate the enzyme, so that their use for precipitation is not recommended in most cases. Instead ammonium sulphate and other salts are used for precipitation in a process called 'salting out'. Salts can change the structure of the solvent, which can lead to large changes in protein conformation by altering the electrostatic interaction between charged groups on the protein surface. The salt also competes with the protein for solvent molecules and thereby lowers its solvation. In large-scale enzyme precipitation, use of many other neutral salts is preferred over ammonium sulphate, which is corrosive and releases NH_3 at higher, temperatures.

Some organic solvents like acetone, methanol and ethanol

are also used for enzyme precipitation, because water miscible solvents decrease the solubility of proteins, leading to precipitation. They are cooled upto 40-60°C before their use, and precipitation is carried out at 0°C, because precipitation at room temperature causes denaturation of the enzyme in most cases. Organic solvents are added drop by drop to avoid local concentration.

Water soluble non-ionic polymers such as polyethylene glycol, alginate, pectinate, carboxymethyl cellulose, polyacrylic and polymetaacrylic acids, etc. also cause enzyme precipitation. Poly-ethyleneimine is also widely used as protein precipitant at large scale. They primarily act through the removal of solvent sphere of the enzyme protein.

Protein's/Enzyme's Purifying Process

Protein purification involves three steps, i.e. dialysis, chromatography and electrophoresis. These three techniques and their variants are briefly described in the following text.

Dialysis : Dialysis is the process that is used to remove small molecules from crude enzyme. For this, enzyme precipitate obtained in the previous step is dissolved in a small quantity of buffer solution, in which the enzyme was originally extracted. The solution is taken in a dialysis bag (may be a cellophane tube) and after sealing securely, the bag is suspended in either distilled water or a buffer of known molarity and ionic composition. Some other salts or chemicals may have to be added sometimes in the outer solution, to prevent the loss of enzyme activity during dialysis. The dialysis is carried out for a few hours with regular change of the outer solution or distilled water (Fig.). At large-scale enzyme purification, dialafiltration' instead of dialysis is used. The enzyme solution is filtered for small molecules through a membrane generally mounted on a fibrous support, by pressure-driven, operation.

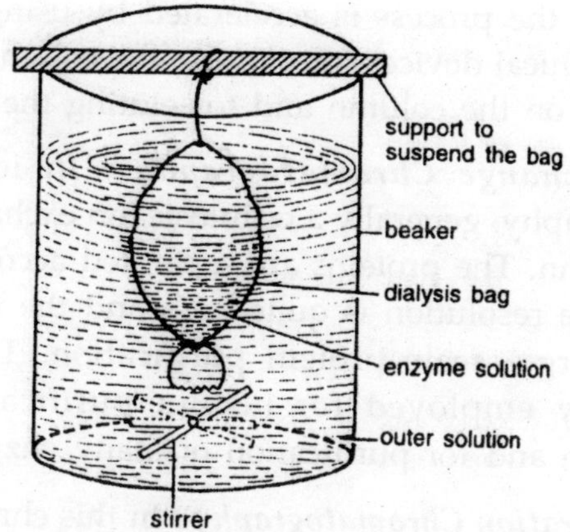

An arrangement for dialysis to purify an enzyme.

Chromatography of Different Types

Chromatographic separation of proteins is the most common method of enzyme purification. Following four types of chromatography are available for this purpose : (i) adsorption or column chromatography; (ii) ion exchange chromatography; (iii) gel filtration chromatography; and (iv) affinity chromatography.

Adsorption Chromatography : In this chromatography, the protein or enzyme solution suspected to contain other proteinaceous impurities, is passed through a column of inert material packed in a glass or steel tube. Most commonly used column materials include finely divided solids such as charcoal, silica, alumina, calcium phosphate, hydroxyapatite, etc. The effluent solution is continuously collected in small fractions of 1.0 to 2.0 ml. The protein in each fraction is estimated by measuring the absorption at 280 nm using a UV spectrophotometer. The enzyme is also assayed in each fraction. Various spleen enzymes such as basic RNAase, acidic RNAase, acidic DNAase, phosphodiesterase, phosphomonoesterase, etc. are often separated from each other using adsorption chromatography. For large-scale chromatographic separation of enzymes, the process is accelerated by using motors and other mechanical devices for packing the column, for loading the enzyme on the column and for eluting the enzyme.

Ion Exchange Chromatography : In ion exchange chromatography, generally a cellulosic ion exchanger is taken in the column. The proteins are separated according to their charges. The resolution is quite high and the technique can facilitate large- scale protein purification. This has been successfully employed for insulin purification, plasma fractionation and for purification of many enzymes.

Gel Filtration Chromatography : In this chromatography, various proteins are separated on the basis of differences in their molecular size. This technique of gel filtration chromatography is also known as molecular exclusion chromatography or molecular sieve chromatography. The basic arrangement for gel filtration chromatography is similar to that for adsorption chromatography. A column made up of glass or steel is taken and packed with a gel (Fig.). The most

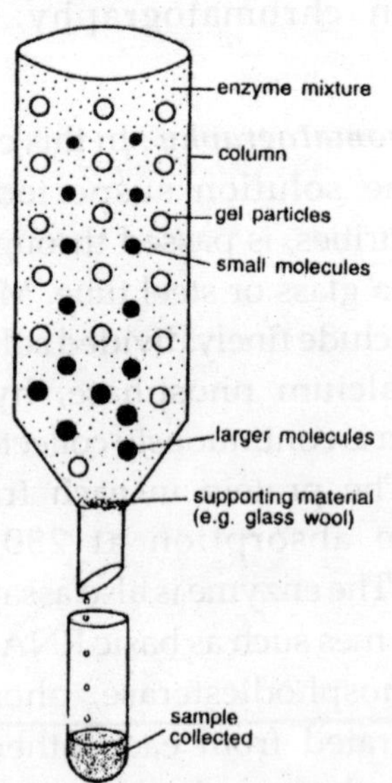

An arrangement for gel filtration chromatography to purify an enzyme.

commonly used gel is 'sephadex', which is a cross linked dextran produced by certain strains of bacteria. Several types of sephadex, namely G-10, G-30, G-50, G-100, G-150, G-200, etc. are available, which differ according to their pore size.

When a mixture of enzymes or proteins is poured on top of the column, different proteins move downwards according to their molecular sizes and come out from the column in order of decreasing sizes; the larger molecules are eluted first. The elution volume is logarithmically proportional to the molecular size.

Gel filtration chromatography can be used for determining the molecular weight of the protein by calibrating the column with proteins of known molecular weights.

Affinity Chromatography : In this method, enzymes are purified according to their specificity for a particular substrate or cofactor. One component of the mixture, containing the enzyme, binds covalently to the solid support of the column, and the other components percolate down through the column. The basic requirements for affinity chromatography are the same as in adsorption or gel filtration chromatography but the packing gel must have some component which can bind with one component of the mixture.

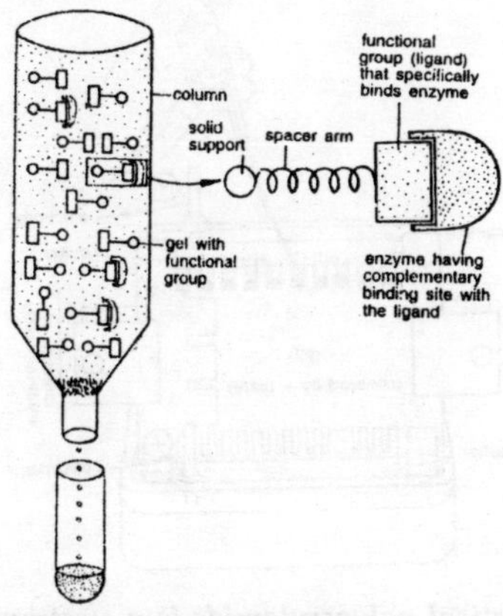

An arrangement for affinity chromatography, showing the role of functional group having affinity with the enzyme.

Many commercial gels available for affinity chromatography contain functional groups attached to the 'spacer arms' of the gel. The 'spacer arm' is a chemical linkage between the functional group and the gel or matrix proper, so that the binding between functional group and the enzyme is kept away from the gel. Thus, the steric hindrance will be unlikely to prevent binding of the specific enzyme to the column.

Technique of Electrophoresis

Electrophoresis is a technique in which molecules (enzymes, proteins, amino acids, nucleotides and nucleic acids) are sepa rated by differences in their net charge in the presence of an externally applied electric field. This technique is routinely used in enzyme purification and isozyme separation in the laboratories, although it has found only limited application at large scale, since the technique is time consuming and is a bit expensive.

Various types of instrumental approaches have been used to separate and purify charged molecules using electrophoresis. However, the most common method for purifying enzymes is through electrophoresis on polyacrylamide gel. Polyacrylamide is a polymer of acrylamide and methylene bisacrylamide, and

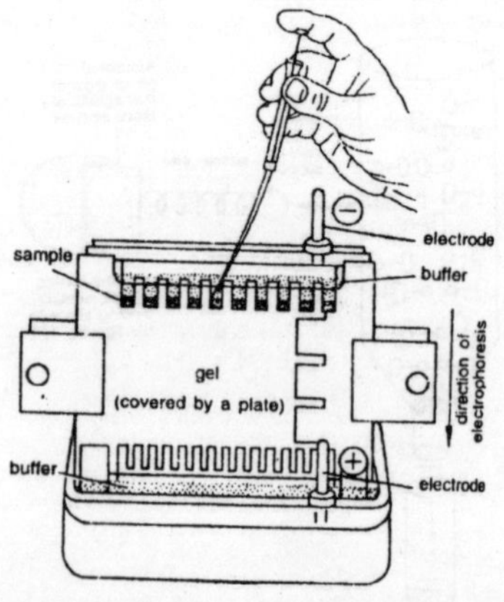

An arrangement for vertical polyacrylamide is a electrophoresis technique for separation of macromolecules.

when prepared as a gel it is transparent, thermostable, non-ionic and extremely regular in structure. The gel may be taken either in the form of a column or a slab, although the latter is preferred over the former (Fig.). The protein mixture is loaded on the gel and the components are separated under a direct current of constant voltage. The migration rate of various components of a mixture is dependent upon their charge and molecular weight.

A variation of the above polyacrylamide gene electrophoresis is the sodium dodecyl sulfate polyacrylamide gel electrophoresis (SDS-PAGE),B which is used to determine the molecular weight of proteins. In this method, the separation is caused by the sieving action of the gel. The proteins migrate through the gel depending on their shapes and *mass* charge (m/z) ratio. Gel electrophoresis is also used to separate various isozymes of a given enzyme: (Isozymes perform the same catalytic function but differ in their regulatory and some kinetic aspects).

One Dimensional SDS-PAGE : SDS-PAGE, is a modified version of the original electrophoresis technique, and was developed in mid-1960's, to facilitate better separation and analysis of proteins. The proteins are not held in an aqueous solution, but are held in a solution, which includes a powerful negatively charged detergent called sodium dodecyl sulphate (SDS). SDS binds to hydrophobic regions of protein molecules and causes them to unfold into extended polypeptide chains. It also releases these chains from their association with other proteins or lipid molecules. Another reducing agent, mercaptoethanol is added to break any S-S linkages, so that subunits of a multi-subunit protein can be separated and analyzed.

The gel used in SDS-PAGE is a cross-linked gel of polyacrylamide in the form of an inert matrix through which proteins migrate. This gel is prepared by polymerization from monomers, just before using. The pore size in this gel can be adjusted to retard the passage of the protein of interest. Through this gel, when a mixture of SDS-solubilized proteins is electrophoresed by applying a voltage, proteins migrate towards the positive electrode. Since native structure of protein is unfolded by SDS and mercaptoethanol, the proteins of same size behave similarly and proteins of different sizes form a series of discrete bands.

The gel can be stained with a dye such as coomassie blue and major proteins can be visualized as stained bands. Minor bands can also be seen by treating the gels with a silver stain. Since polypeptides are separated due to size and since subunits are also separated, molecular weights and subunit composition of a protein complex can be determined by SDS-PAGE. Membrane proteins, cytoskeleton proteins, seed storage proteins, etc. have been successfully analyzed using SDS-PAGE during 1970's and 1980's.

Two-dimensional PAGE : In one-dimensional electrophoresis, bands sometimes overlap, so that only a limited number of proteins (upto 50) can be resolved (Fig), but if after completing electrophoresis in one direction by a method called isoelectric focusing (Fig), the proteins are separated further in another direction (at right angle to the first direction and in presence of SDS), as many as 2000 proteins can be separated as a two- dimensional protein map (Fig.).

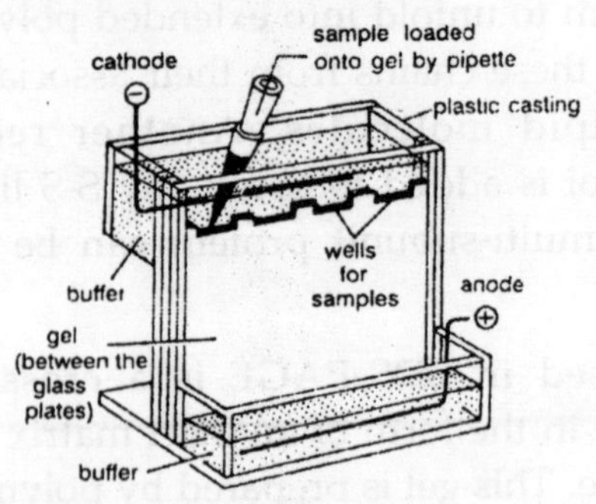

(a) apparatus for SDS - PAGE

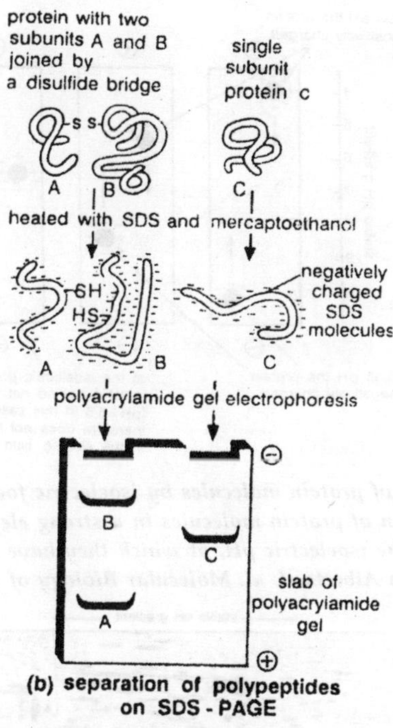

Apparatus for PAGE (a) and separation of molecules through PAGE using SDS (b).

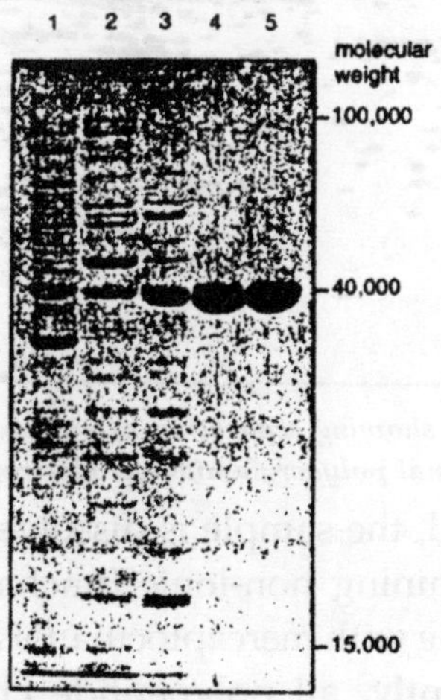

Photograph of a gel after SDS-PAGE; the lanes from left to right show successive stages of purification of an enzyme; note that protein bands gradually disappear during purification.

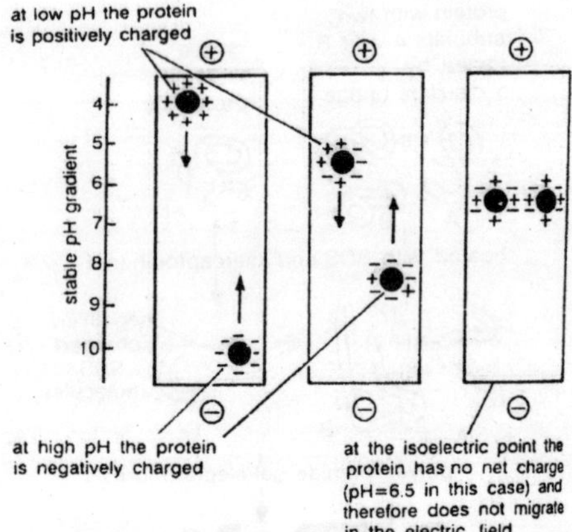

Separation of protein molecules by isoelectric focusing notice the migration of protein molecules in a strong electric field till they reach the isoelectric pH, at which they have no net charge (redrawn from Alberts el. al. Molecular Biology of the Cell, 1994).

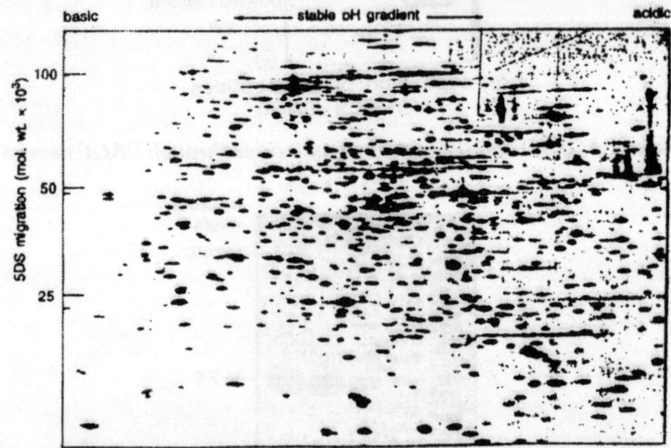

Photograph of a gel showing separation of proteins of E. coli. cells, using two dimensional polyacrylamide-gel electrophoresis (PAGE).

In this method, the sample is dissolved in a small volume of a solvent containing non-ionic (uncharged) detergent (no SDS is added) along with mercaptocthanol and urea (denaturing agent). Consequently, all polypeptide chains are denatured and dissociated, but the intrinsic charge remains unchanged (in SDS-PAGE the intrinsic negative charge is enhanced). The polypeptide chains are now separated by a procedure called isoelectric focusing, which depends on the principle that the charge of the protein varies with pH of the solution (at high pH it is negatively charged and at low pH it is positively charged). Furthermore, at a specific pH (called isoelectric point), a particular protein has no net charge, so that at this point it will fail to migrate in an electric field.

Thus for isoelectric focusing, proteins are electrophoresed in a narrow tube with polyacrylamide gel having a pH gradient established by a mixture of buffers. Each protein migrates to its own isoelectric point and stays there giving first dimension of the two-dimensional PAGE. If this narrow gel, containing proteins already separated in one dimension, is again subjected to electrophoresis in a direction at right angle to the direction of the first separation, upto 2000 polypeptide chains can be resolved. While subjecting to electrophoresis in the second dimension, the narrow gel with proteins already separated in one dimension is soaked in SDS and placed on one edge of an SDS-PAGE-gel slab, through which each polypeptide chain then migrates and forms a discrete spot.

Only those proteins can remain unresolved, which have identical size and identical isoelectric point, a situation, which is rather rare. Staining procedures or autoradiography can be used for detection of these resolved proteins, so that even proteins that differ by a single amino acid can be resolved. Antibodies coupled with radioisotopes, specific enzymes or fluorescent dyes can be used for detection of specific proteins, following a procedure called western blotting or immunoblotting.

Protein's Identification

In the early years of twentieth century (1901-1907), Emil Fischer discovered that a protein consists of amino acids and that a few amino acids can be condensed to obtain a small chain, which he called peptide. We also know that a protein molecule can be broken down into its amino acids by heating it with an acid. In the mixture of amino acids thus obtained, one can find out the composition of amino acids, i.e. how much of each amino acid is present in this mixture.

These amino acids could be neatly separated by starch column chromatography, drop by drop in separate containers. The solution in each container is treated with a chemical that turns the amino acid into a coloured compound. The intensity of colour in this compound is a measure of the amount of a particular amino acid present in the solution. This intensity of colour is measured by an instrument called spectrophotometer, which is based on the principle that the amount of substance is directly proportional to the amount of light of particular wavelength absorbed by the substance. Later 'two-dimensional paper chromatography' and 'gas chromatography' were developed and used for study of amino acid composition of proteins.

Erwin Brand suggested a system of symbols for the amino acids, where each amino acid was represented by the first three letters (with some exceptions) of the amino acid. Consequently, after knowing the composition of the blood protein called 'serum albumin', its, 526 amino acids could be represented as Gly15 Val45 Leu58 Ile9 Pro31 Phe33 Tyrig Trpi Ser_{22} Thr_{27} Cys_{32} $CySH_4$ Met_6 Arg_{25} His16 Lys58 Asp46 Glugo- Later, single letter abbreviations (alphabets) have bee assigned to each amino acid, so that the composition of serum albumin could be written as G15 V45 L58 I_9 P_{31} F_{33} $Y_{'8}$ W, S_{22} T_{27} C_{32} Csh_4 M_6 R_{25} H_{16} L_{58} N_{46} Ego. However, this formula indicated only the composition, and did not indicate in what sequence the 526 amino acids of 'serum albumin' were arranged. This problem was later tackled and solved (see below).

Determining of Amino Acid Sequence of a Protein : In 1945, Frederick Sanger in UK developed a technique to determine the sequence of amino acids in a peptide chain. For this purpose, he selected 'insulin', which was important and was small in size to be used in such an exercise. Sanger started by trying to identify the amino acid at the amine end of the chain (N-terminal amino acid). Since the amino group of this amino acid is supposed to be free, Sanger used a labelled chemical that combines with free amino (-NH_2) group, but not with a peptide bond (-CO-NH-). This produces a DNP (dinitrophenyl) derivative of the peptide chain, which has the N-terminal amino acid labelled with DNP. Sanger could break the chain into its individual amino acids and isolate the one with DNA label (due to yellow colour of DNP on paper chromatography) and identify it through techniques available for identification of amino acids. The amino acid at C-terminal end could also be similarly isolated and identified. By removing amino acids one by one, sequence of several amino acids at each end could be determined.

Using the above approach, Sanger and Hans Tuppy (a co-worker) determined the complete amino acid (aa) sequence of the two peptide chains (G-chain with 21 aa and P-chain with 30 aa), which the protein molecule of "insulin" contains. The two peptide chains were first separated and each fragment in its turn broken (partial hydrolysis) into fragments, each consisting of two to three amino acids, in which N-terminal end could be identified by DPN labelling.

In dipeptide now, if the N-terminal amino acid is known, the other will be obtained by further digestion and will be assigned the C-terminal position in this dipeptide. This approach was successfully utilized later for determining the complete amino acid sequences of G-chain and P-chain of insulin. This work was completed in 1952. Later, they could also determine how the two chains were joined together, so that in 1953 complete sequence of the first protein (insulin) was announced, for which Sanger was awarded Nobel Prize in Chemistry for the year 1958.

Synthesizing First Protein (oxytocin) in the Laboratory : Since the amino acid sequence of a polypeptide could be known, attempts were also made to synthesize a protein with a pre-determined order of amino acids. The first protein that was synthesized in the laboratory was "oxytocin", which was a hormone made up of only 8 amino acids and performing important functions in the body. In 1953 Vincent du Vigneaud successfully synthesized a protein molecule that represented oxytocin, and exhibited all the properties of the natural hormone. Du Vigneaud was awarded Chemistry Nobel Prize for the year 1955.

Determining Process of Three-dimensional Structure of Protein : Although, primary structure of a protein could be determined and used for synthesis of proteins, the three-dimensional (TD) structure of a protein, showing how each amino acid is bent or curved was more difficult. Max Ferdinand Perutz took haemoglobin (a blood protein) and John Cowdery Kendrew used myoglobin (a muscle protein) for X-ray diffraction studies to resolve their TD; structures. Using this technique, in 1959, the TD structure of myoglobin and in 1960, the TD structures of haemoglobin were successfully determined. Both Perutz and Kendrew shared the Nobel Prize in Chemistry for the year 1962. Besides x-ray crystallography, nuclear magnetic resonance (NMR) is also used for the study of TD structure of proteins. Both these techniques are briefly described below.

X-ray Diffraction and NMR Spectroscopy : Amino acid sequences in proteins and nucleotide sequences in nucleic acids can sometimes be used to predict the secondary structures of these molecules. For instance, helix structure of a transmembrane segment of a protein can be predicted from its amino acid sequence. Similarly, the clover leaf model of tRNA was predicted from the nucleotide sequences of a number of tRNA molecules. But these sequences did not allow us to predict the three-dimensional structures of these macromolecules. X-ray diffraction and nuclear magnetic resonance (NMR) spectroscopy have been used extensively for the study of three-dimensional structures of cellular macromolecules, particularly the proteins and nucleic acids.

In x-ray crystallography, the macromolecule should be available in a crystal form and x-rays (wave length=0.1 nm = diameter of hydrogen atom which is much smaller than other electromagnetic radiations) are allowed to pass through crystal sample. Most of the x-rays pass through, but small fraction is scattered by the atoms in the sample. If these atoms are well ordered, the diffracted x-rays may reinforce and make diffraction spots to be recorded by a detector. By looking at the spots in the x-ray diffraction patterns, information about the positions of atoms in the crystal can be deduced. From this information (available in the form of a complex three-dimensional electron density map) and the information about the amino acid/nucleotide sequences of a macromolecule, three-dimensional structure can be deduced using a computer device.

NMR spectroscopy (proposed by Jeener in early 1970's) has been used largely to analyze the structure of small molecules like small proteins or protein domains. It does not need the sample in crystalline form, but only in the form of a solution that is placed in a strong magnetic field. Certain atomic nuclei, particularly hydrogen nuclei, have a magnetic moment or a spin (which means intrinsic magnetization, like a bar magnet), so that they align along the magnetic field in which they are placed. However, if radio-frequency (RF) pulses of electromagnetic radiation are applied from outside, the spin of atomic nuclei, which remains aligned to the surrounding magnetic field, can be changed to a misaligned excited state. These misaligned excited nuclei, when relaxed to their aligned state, they emit RF radiation, which can be measured and displayed as a spectrum.

The nature of these emitted radiations depends on the environment of each hydrogen nucleus, so that if one hydrogen nucleus is excited, it will influence the absorption or emission of radiation by other nuclei in the neighbourhood. Therefore, the signals from hydrogen nuclei in different amino acid residues will differ. The shift in these signals can be measured and used to obtain information about distances between different parts of a small protein molecule. When this information is used with the information about amino acid sequences, three-dimensional structures can be obtained for molecules of 15,000-20,000 daltons molecular weight. In future, molecules with molecular weights of upto 30,000-40,000 daltons will hopefully be analyzed by NMR, but bigger molecules cannot be examined by this method. Many functional domains of proteins and complex carbohydrate side chains of glycoproteins have also been examined using NMR spectroscopy.

Bibliography

Ahmad, K. Henikoff, S. : *Epigenetic Consequences of Nucleosome Dynamics,* MIT Press, Cambridge, 2002.

Altaian, S. : *Transfer RNA,* MIT Press, Cambridge, 1978.

Anderson, S. : *Sequence and Organization of the Human Mitochondrial Genome Nature,* Oxford University Press, London, 1981.

Andre, C. A. Levy and V. Walbot : *Structure of Plant Mitochondrial Genomes,* John Wiley, London, 1992.

Atkins, J. F. : *Recoding : Dynamic Reprogramming of Translation,* Oxford, New York, 1996.

Bennett, R. H. : *Transcription, Processing and Editing in Plant Mitochondria,* Oxford, New York, 1992.

Bloom, L. B. : *Biochemical Basis of DNA Replication Fidelity,* MacMillan, London, 1993.

Branden, C. and J. Tooze : *Introduction to Protein Structure,* Garland, New York, 1991.

Darnell, J. E., H. F. Lodish and D. Baltimore : *Molecular Cell Biology,* Scientific American Books, New York, 1995.

Das, A. : *Control of Transcription Termination,* Routledge, New York, 1993.

Davidson. J. N. : *Biochemistry of Nucleic Acids,* Cox and Wyman, London, 1965.

Demidov, V. V. : *PNA and LNA throw Light on DNA, Trends,* Oxford, New York, 2003.

DuPraw, E. J. : *Cell and Molecular Biology,* Academic Press, New York, 1969.

Ebby, S. R. : *Non Coding RNA Genes and Modern RNA,* Oxford, New York, 2001

Echols, H. : *Fidelity Mechanisms in DNA Replication,* Oxford, New York, 1991.

Ellis, R. J. : *Chloroplast Protein Synthesis,* Oxford, New York, 1975.

———— : *Molecular Chaperones,* Oxford, New York, 1991.

———— : *The Biogenesis of Chloroplasts Protein,* Oxford, New York, 1974.

Fischer, L.M. : *DNA Unwinding in Transcription and Recombination,* Oxford, New York, 1982.

Friedberg, E. C. : *Relationships Between DNA Repair and Transcription,* Oxford, New York, 1996.

Gesteland, R. F. and J. F. Atkins : *The RNA World,* Oxford, New York, 1993.

Gilbert, W. and B. Muller : *Isolation of the Lac Repressor,* Oxford, New York, 1966.

Gilham, N. W. : *Genetic Analysis of the Chloroplast and Mitochondrial Genomes,* Oxford, New York, 1974.

Gilmour, R. S. and J. Paul : *Tissue Specific Transcription of the Globin Gene in Isolated Chromatin,* Oxford, New York, 1973.

Glick, B. and G. Schatz : *Import of Proteins into Mitochondria,* Oxford, New York, 1991.

Gray, M. W. : *The Evolutionary Origin of Organelles,* Oxford, New York, 1989.

Grunstein, M. : *Histones as Regulators of Genes,* Oxford, New York, 1992.

Gupta, P. K. : *Cell and Molecular Biology,* Rastogi Publications, Meerut, 2000.

Hadjiolov, A. A. : *The Nucleolus and Ribosome Biogenesis,* Oxford, New York, 1985.

Hoagland, M. G. : *The Present Status of the Adaptor Hypothesis,* MacMillan, London, 1959.

Hoeijmakers, J. H. : *Nucleotide Excision Repair from Yeast to Mammals,* MacMillan, London, 2001.

Holley, R. W. : *Structure of a Ribonucleic Acid,* McGraw Hill, London, 2002.

Ingram, V. : *The Biosynthesis of Macromolecules,* Benjamin, New York, 2001.

Jacob, F. : *Genetic Regulatory Mechanisms in the Synthesis of Proteins,* Oxford, New York, 1961.

Jukes, T. H. : *Mitochondrial Codes and Evolution,* Oxford, New York, 2000.

Kaguni, L. S. : *Eukaryotic DNA Polymerase-primase : Structure, Mechanism and Function,* MacMillan, London, 1999.

Kaziro, Y: *Structure and Function of Signal Transducing GTP Binding Proteins,* Blackwell, London, 2001.

Kendrew, J. C. : *The Three-dimensional Structure of a Protein Molecule,* Sinauer, New York, 1991

Kornberg, A. and Baker, T.A. : *DNA Replication,* Freeman, New York, 2002.

Kramer, A. :*The Structure and Function of Proteins Involved in Mammalian Pre-mRNA Splicing,* Grandada, New York, 2001.

Kurland, C. G. : *Structure and Function of the Bacterial Ribosomes,* Oxford, New York, 2000.

Lake, J. A. : *The Ribosome,* Oxford, New York, 1981.

Lehninger, A. L. : *The Mitochondrion,* Academic Press, New York, 2001

Lewin, Benjamin : *Genes,* Oxford University Press, New York, 2000.

Maxwell, E. S. and Fournier, M.J. : *The Small Nucleolar RNAs,* MacMillan, London, 1995.

McKnight, S. L. : *Molecular Zippers in Gene Regulation,* Princeton, New York, 2002.

Merrick, W. C. : *Overview : Mechanism of Translation Initiation in Eukaryotes,* Longman, London, 2001.

Moore, P. B. : *Molecular Mimicry in Protein Synthesis,* Cambridge, London, 2002.

Noller, H. F. : *Ribosomol RNA and Translation,* Oxford, New York, 1999.

Olson, R. E. : *Enzymatic Synthesis of DNA,* John Wiley, New York, 2001.

Owen, L. :*DNA Synthesis,* Freeeman, San Francisco, 2001.

Pabo, C. O. and R.T. Sauer : *Transcription Factors : Structural Families and Principles of DNA Recognition,* Printice Hall, London, 1993.

Pardee, A. B. : *Molecular Basis of Biological Regulation: Origins of Feedback Inhibition and Allostery,* Arnold, London, 2003.

Pardue, M. L. and J. G. Gall : *Molecular Hybridization of Radioactive DNA to the DNA of Cytological Preparations,* Dekker, New York, 2001.

Paszkowski, J. : *DNA and Histone Methylation in Plants,* MacMillan, London, 2002.

Razin, A. and A. D. Riggs : *DNA Methylation and Gene Function,* Longman, New York, 2002.

Rich, A. and S. H. Kim : *The Three Dimensional Structure of Transfer RNA,* Oxford, New York, 1999.

Robertson, M. : *RNA Processing : the Post-RNA World,* Oxford, New York, 1999.

Rosenberg, M. and D. Court : *Regulatory Sequences Involved in the Promotion and Termination of RNA Transcription,* Cambridge, New York, 2003.

Sager, R. : *Cytoplasmic Genes and Organelles,* Academic Press, New York, 2003.

Schmid, M. B. : *Structure and Function of the Bacterial Chromosome,* Granada, London, 1990.

Spector, D. L. : *The Dynamics of Chromosome Organization and Gene Regulation,* Oxford, New York, 2005.

Stryer, L. : *Biochemistry,* W.H. Freeman & Co., New York, 2002.

Temin, H. M. : *RNA Directed DNA Synthesis,* MacGraw-Hill, New York, 2002.

Turner, B. M. : *Histone Acetylation and Control of Gene Expression,* MacMillan, London, 2002.

Ulrich, B. : *Mechanism and Regulation of Eukaryotic Protein Synthesis,* Oxford, New York, 2002.

Uptain, S. M. : *Basic Mechanisms of Transcript Elongation and its Regulation,* MacMillan, London, 1997.

Vernadsky, V. I. : *Organellar Heredity,* Revan Press, New York, 1978.

Voinnet, O. : *RNA Silencing : Small RNAs as Ubiquitous Regulators of Gene Expression,* MacMillan, London, 2002.

Wallace, D. C. : *Diseases of the Mitochondrial DNA,* Blackwell, New York, 2002.

Wang, T. S. F. : *Eukaryotic DNA Polymerases,* MacMillan, London, 2002.

Watson, J. D. : *Molecular Biology of the Gene,* Oxford, New York, 2002.

———— : *Molecular Structure of Nucleic Acids : a Structure for Deoxyribose Nucleic Acid,* Oxford, New York, 2001.

Wilkins, M. H. F. : *Physical Studies of the Molecular Structure of Deoxyribose Nucleic Acid and Nucleoprotein,* MacMillan, London, 2002.

Williams, R. C. : *Reconstitution of Active Tobacco Mosaic Virus from its Inactive Protein and Nucleic Acid Components,* Oxford, New York, 1955.

Williamson, B. : *Split Gene Transcription,* Oxford, New York, 2001.

Woese, C. R. : *The Genetic Code : The Molecular Basis for Genetic Expression,* Harper & Row, New York, 2003.

Wuthrich, K. : *Protein Structure Determination in Solution by Nuclear Magnetic Resonance Spectroscopy,* Oxford, New York, 2001.

Yamamoto, T. : *A Bipartite DNA Binding Domain Composed of Direct Repeats in the TATA Box Binding Factor TFHD,* Oxford, New York, 2001.

Yamashita, S. : *Transcription Factor TFHB Sites Important for Interaction with Promoter-bound TFIID,* Springer, London, 2000.

Yams, M. : *Translational Efficiency of Transfer RNAs : Uses of an Extended Anticodon,* MacMillan, London, 2003.

Yanofsky, C. : *The Tryptophan Synthetase System,* Oxford, New York, 2001.

Zawel L., and D. Reinbcrg : *Common Themes in Assembly and Function of Eukaryotic Transcription Complexes,* MacMillan, London, 2002.

Zuckerman, B. : *Initiation of mRNA Translation in Prokaryotes,* Oxford, New York, 1990.

Zurick, D. : *Origin and Evolution of Mitochondrial DNA,* Oxford, New York, 1989.